Herausgeber: Markus Greim, Wolfgang Kusterle und Oliver Teubert

Rheologische Messungen an Baustoffen 2020

Tagungsband zum 29. Workshop und Kolloquium, 11. und 12. März an der OTH Regensburg

ISBN:

978-3-347-02886-9 (Paperback)

978-3-347-02887-6 (Hardcover)

978-3-347-02888-3 (e-Book)

1. Auflage 2020

Layout: Christian Greim

Satz: Dr. Helena Keller, Christian Greim

Verlag und Druck: tredition GmbH; Halenreie 40-44; 22359 Hamburg; www.tredition.de

Publisher: Markus Greim, Wolgang Kusterle and Oliver Teubert

Rheological Measurement of Building Materials 2020

Proceedings of the 29th Conferences and Laboratory Workshops, 11th and 12th March at OTH Regensburg

ISBN:

978-3-347-02886-9 (Paperback)

978-3-347-02887-6 (Hardcover)

978-3-347-02888-3 (e-Book)

1. Edition 2020

Layout: Christian Greim

Typesetting: Dr. Helena Keller, Christian Greim

Published and Printed by: tredition GmbH; Halenreie 40-44; 22359 Hamburg; www.tredition.de

Preface

The art of rheological measurements for construction materials was until 50 years ago a totally unknown territory. Even today it is often hard to understand, unless you are an expert in the field. There are many people who buy the costly equipment for rheological measurements. However, they often find after some time that the test set up, numerous influencing factors and the proper interpretation of the results is not that easy to work out. As a consequence this expensive equipment ends up lying in the corners of the labs gathering dust.

This conference and workshop on the "Rheology of Building Materials" in Regensburg cannot solve all these problems. It, however, supplies a platform for excellent presentations of recent research work as well as demonstrations of useful equipment and exchange of ideas.

The proper workability of materials bound by mineral binders is essential for an economic production and for a perfect performance in hardened state. The papers in these proceedings show examples of enhanced testing and applications.

These proceedings address important rheological topics:

- Inline control of ready mix concrete
- Tests on concrete and mortar for different purposes (e.g. rheology of calcium-sulfoaluminate mortars, of alkali activated binders, of mortars for 3D-printing)
- The influence of fibres on injecting mortars
- Oscillatory measurements.

The papers on mineral materials are enhanced with a special contribution on the behaviour of granular frazil ice. Hopefully, the reader will enjoy these papers and get a lot of inspiration as well as some new ideas for her/his own work.

Personally, due to my retirement this summer, this conference is the last I chair. In the future I will join you as an ordinary delegate. I wish Markus Greim, Oliver Teubert, Helena Keller from Schleibinger Testing Systems and my still unknown successor as chair a lot of success for the future. I would like to also thank them all for their engagement in this Conference, which is indeed the 29th.

Furthermore, I would like to thank all the authors and conference delegates for their contributions. I hope to meet you next year at the 30th anniversary at the OTH Regensburg.

Wolfgang Kusterle

Regensburg, March 2020

Rheological properties of calcium-sulfoaluminate cement mortars

Jacek Gołaszewski, Małgorzata Gołaszewska
Silesian University of Technology, Gliwice, Poland

Abstract

The paper presents research into the rheological properties of calcium-sulfoaluminate (CSA) cement mortars. CSA cements may prove to be an alternative to Portland cement, due to its lower carbon footprint. High cost of CSA cement is, however, one of detriments to its more widespread use. To mitigate this issue, it may be possible to mix CSA cement with other cements or non-clinker main constituents of cements. Presented research dealt with the issue of rheological properties of mortars with cement which are a mix of CSA cements with 10, 20 and 30% of Portland cement CEM I 42.5R NA and 10, 20 and 30 % of limestone. Tested was early shrinkage of the mortars, the yield stress and plastic viscosity, and an increase of torque in the first 1,5h of hydration. The results indicated that there is no negative interaction between CSA cement and limestone in terms of rheological properties, achieving slightly higher shrinkage than CSA cement, while the effect on yield stress and torque was small, but dependant on the amount of limestone. Addition of CEM I 42.5R NA to CSA cement indicated a negative interaction between two binders in terms of yield stress and torque changes in time, however the effect on shrinkage was small, but not inherently negative as the shrinkage was significantly lower than in case of Portland cement CEM I 42.5R NA.

1 Introduction

Calcium sulfoaluminate (CSA) cements have been gathering more interest in the last few years, due to the lesser impact of their production on environment in comparison to ordinary Portland cement (OPC) [1,2] In case of OPC, for each ton of clinker produced, on average 0.85 tons of CO_2 is produced, mostly due to the chemical process of calcination of limestone and high heat needed to obtain required clinker phases (~1450°C) [3,4]. In case of CSA, the emissions of CO_2 are lower, due to lower temperature in kiln (~1250°C), lower amount of limestone calcinated during production, as well as less energy needed to grind the obtained CSA clinker [1,5,6]. This translates into lower emission of CO_2 in comparison to OPC; research of Li et al [7]indicated that the decrease in emission of CO_2 is in range of 20-30%, while Ellis [8] estimated, that there can be up to 35% of reduction in CO_2 emissions.

CSA cement is a mineral hydraulic binder, which clinker is obtained by burning limestone, bauxite (or other aluminum-rich rocks) and gypsum, however there is research being conducted on using waste materials for CSA production [9–11]. Main mineral phases of sulfoaluminate (SA) cement is ye'elemite ($C_4A_3\hat{S}$) and belite (C_2S) with secondary phases of calcium sulphate ($C\hat{S}$), and small amounts of aluminoferrite (C_4AF). Completely different phase composition is responsible for different hydration of CSA cements. The main product of the reaction of hydration of CSA cement is ettringite $C_6A\hat{S}_3H_{32}$, $C_2A\hat{S}H_8$ (stratlingite) and monosulphate, with CSH phases appearing after 14-30 days of the hydration start [12,13]. This leads to many beneficial characteristics of CSA cement: it has a short initial setting time, fast strength development and low shrinkage.

CSA cements have been introduced by A. Klein in the 1960sin the USA, and decade later developed mostly by Chinese enterprises [14]. Due to its short setting time, high early strength and low shrinkage, they found use in production of prefabricated elements, bridges, and shotcrete. [15] The scope of using CSA cements is limited, mostly due to the high production cost connected to the rarity of raw materials (especially bauxite), therefore the research in the field is often directed to mixing CSA cement with other materials to obtain binders of similar characteristics but of a lower cost.

American Guide for the Use of Shrinkage-Compensating Concrete [16] has even included the mixture of Portland cement and CSA cement in amount of 10-30% as a cement type "K" of expansive cements. Chaunsali and Mondal [17] had found that mixing OPC with 15% CSA cement caused a significant expansion of cement paste, however Le Saoût et al. [18] had shown, that the chemical shrinkage was higher in case of pastes with 90% OPC and 10% of CSA as a binder than in OPC. There was also research conducted by Huang et al. [19] into the rheological properties of fresh mortars with CSA and OPC blends, which had shown that the addition of CSA to OPC may increase the yield stress and viscosity of the mortar, depending on the anhydrate content. It must be noted, however, that the topic of rheology of CSA – OPC blends was not well developed in existing literature.

Similarly, there is very little information concerning the rheology of the CSA blends with mineral fillers, such as limestone. There are studies available that show the positive effect of limestone of hydration of CSa, mostly due to the stabilization of ettringite in the presence of hemicarboaluminate and monocarboaluminate which can appear in the presence of limestone. [10,20–22]

Thus, the aim of the presented study was to investigate the rheological properties of mortars with CSA blends with CEM I 42,5R and limestone in amount 10, 20, and 30% of binder mass. Also tested was early shrinkage measured by the cone method. Rheological parameters of yield stress and plastic viscosity were also tested, as well as the change of torque in first 1h of hydration.

2 Materials and methods

Two types of commercially available cements were used in the research: CSA cement and CEM I 42,5R NA cement, which chemical and phase composition are shown in Table 1 and Table 2 respectively. One type of limestone was used in the research – commercially available ground limestone, which complies with the requirements set by standard EN 197-1 [23] and EN 197-2 [24] for limestone used for as a main constituent of cements The chemical composition of limestone is shown in Table 1.

The specific surface area of CSA cement was 4500 cm^2/g, CEM I 42,5R NA cement had SSA of 3950 cm^2/g and limestone - 4800 cm^2/g. The initial setting time determined according to standard EN 196-3: 2016 [25] of used CSA cement to was 20 minutes, while CEM I 42.5R NA was 260 min.

Table 1: Chemical composition of materials used in the research

Cement type	Constituent [%]								
	LOI	SiO_2	Al_2O_3	Fe_2O_3	CaO	MgO	SO_3	Na_2O	K_2O	Na_2O_{eq}
CSA	0.46	9.2	28.1	1.52	39.2	3.5	11.4	0.08	0.35	-
CEM I 42,5R NA	2.8	20.55	4.67	2.8	64.35	1.18	2.79	0.18	0.43	0.46
Limestone MW	42.7	1.4	0.4	0.5	53.2	1.5	0.02	-	-	-

Table 2: Phase composition of cements used in the research

Cement type	Constituent [%]							
	C_3S	C_2S	C_3A	C_4AF	$C_4A_3\hat{S}$	$C\hat{S}$	$3C_2S*3C\hat{S}*CaF_2$	MgO
CSA	-	10.4	-	1.2	64.9	5	9.4	4.9
CEM I 42,5R NA	62.4	12.2	7.6	8.5	-	~2	-	-

Where: C – CaO, S – SiO_2, A – Al_2O_3, $\hat{S}$ – SO_3, M – MgO

All tests were conducted on mortars, which composition was based on standard mortar according to EN-197 – 1 [23], namely: 450 g of cement, 1350 g of standard sand, and water-cement ratio of 0.5. Due to technical aspects, w/c ratio was raised to 0.6 in case of rheometric tests in anticipation of high yield stress of mortars with CSA cement. Mixing procedure was assumed also according to EN-197 – 1 [23], however in case of CSA cements with 20% and 30% of CEM I 42.5R NA addition, the procedure was shortened due to a rapid workability loss of the fresh mortar, and lasted just 90 s instead of 240 s.

Mortars have been subjected to rheological tests in a rheometer Viskomat NT after 5 minutes from mixing, and then underwent a 1.5 h measurement cycle, shown in Figure 1, which aim was to check the changes in the torque in time. The first five minutes of the cycle was devoted to measuring the yield stress and plastic viscosity of the mortars and thus the rotational speed increased rapidly, and then gradually decreased, while the rest of the measurement cycle had a constant rotational speed.

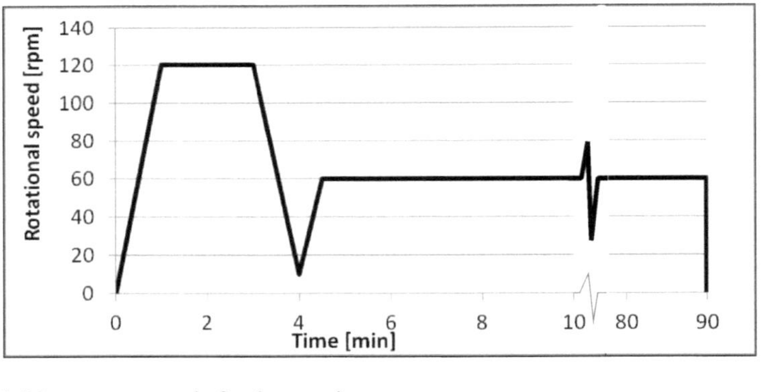

Figure 1: Measurement cycle for rheometric tests.

To calculate the yield stress and plastic viscosity of the mortar, simplified Bingham model was adopted.

$$M = g + hN$$

Where: M – torque, N – rotational speed of measuring cylinder, g – shear resistance, *h* – plastic flow resistance. In this equation, the shear resistance *g* corresponds to the yield stress τ_0 and the plastic flow resistance *h* - plastic viscosity ηp, and they will be referred as such further.

In the research, early shrinkage was measured in Shrinkage Cone deltaEL, as shown in Figure 2. System allows for continuous measurement of fresh mortar, by laser measurement of changes in height of conical sample. The linear change in sample height corresponds to the volumetric shrinkage of the sample. Standard mortars were prepared and placed in the cone immediately after mixing. The measurement itself lasted 24 h. The apparatus was kept in a climatic chamber, in set temperature 20°C and humidity of 65%, for the duration of the tests, to minimize the effects of environmental conditions on the sample.

Figure 2: Shrinkage measurement equipment - Shrinkage Cone deltaEL setup.

3 Results and discussion

3.1 Rheology of fresh mortar

The results of continuous measurement of torque of CSA cement with CEM I 42.5R NA addition are shown in figure 3.

As it can be seen, CSA cement is characterized by the lowest torque, which changed very little in time, rising from around 14 Nmm to 20 Nmm during the first 90 minutes. The torque of Portland cement CEM I 42.5R NA underwent more visible change, increasing from 25 Nmm to over 40 Nmm. However, the mortars with a mix of CSA cement with 10, 20 and 30% of CEM I 42.5R NA had exhibited unusual behavior. The torque of those cements substantially increased, and quickly reached 100 Nmm. The tests have been stopped after the torque reached 100 Nmm, due to the fact, that for the high torque values, mortar can be so stiff, that it does not let the probe pass. This leads to the mortar skidding on the walls of measuring container, and yields wring test results. The increase in torque, caused by rapid loss of workability, made it impossible to accurately measure yield stress and plastic viscosity, thus for mortars with cements with 20% and 30% of CEM I 42.5R NA addition, the measuring process shown in figure 1 was changed to constant rotational speed of 60 rpm. Still, the loss of workability was extremely swift, and the measurement had to be stopped after 10 minutes in case of 30% addition of CEM I 42.5R NA to CSA cement, and was practically impossible for mortars with cement with 20% of CEM I 42.5R NA mass.

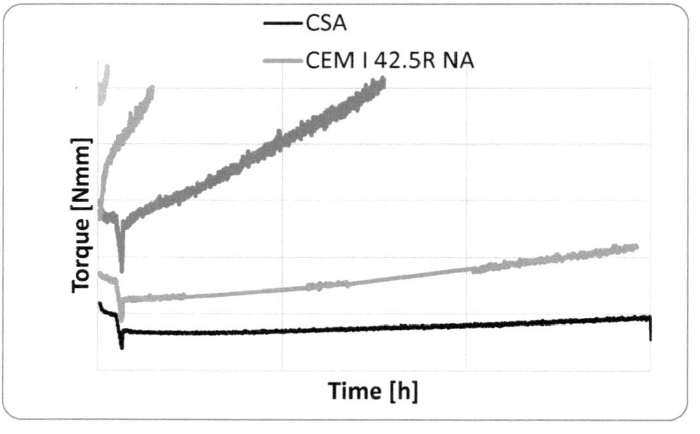

Figure 3: Torque changes in time, for CSA cements with CEM I 45.5R NA addition.

The clear negative effect of adding CEM I 42.5R NA to CSA cement in terms of consistency changes in time has been observed. The negative interaction of CSA and Portland cement CEM I may be caused by overlapping needs of both cements in the beginning of the hydration. In the first minutes of hydration, alumina phases (C_3A) in Portland clinker react with calcium sulphate (gypsum or anhydrite) to create ettringite – as when the amount of gypsum is inadequate, the rapid setting (called "flash set") occurs, as calcium aluminate hydrates are produced [26]. In case of CSA cement, the setting occurs on the basis of ye'elemite reaction with calcium sulphate (usually in form of anhydrite) and water resulting in formation of ettringite [5,10,13]. It may be possible, that the reaction of ye'elemite binds available calcium sulphate, causing the rapid set of C_3A in Portland clinker, causing flash set to occur. The other effect which could be connected to the rapid loss of workability in CSA cements with CEM I 42.5R NA addition is the increase in speed of ye'elemite hydration in presence of portlandite, which is a product of alite (C_3S) hydration [27,28]. This effect requires, however, further consideration and research.

It should be also noted, that in the case of CSA cement, the measurement process significantly delayed the initial setting time, which was determined to be 20 min. The hydration of CSA cement was clearly affected by the continuous mixing. This effect also requires further research.

The results of continuous measurement of torque of mortars with CSA cement with limestone addition are shown in figure 4.

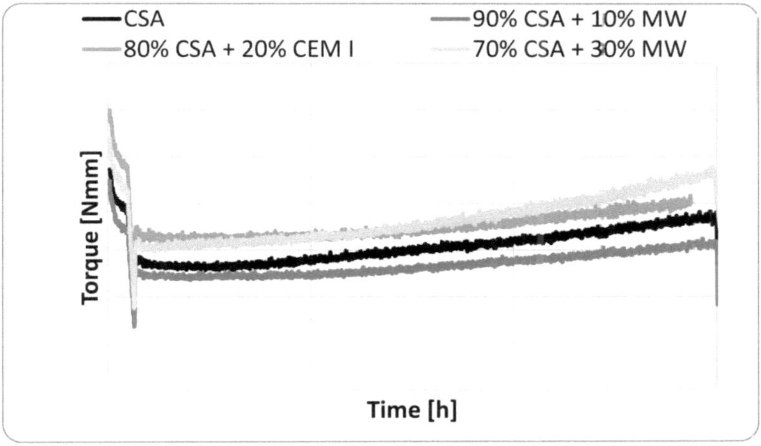

Figure 4: Torque changes in time, for CSA cements limestone MW addition.

There is no clear negative interaction between CSA cement and limestone, as has been observed in the case of CSA with Portland cement CEM I. The addition of lime to CSA cement in amount of 10% of cement mass cement has slightly decreased the torque, while an addition of 20% and 30% of limestone slightly increased the torque, as well as the rate of torque increase. This effect could be attributed to both physical and chemical effect of limestone on CSA hydration.

While the limestone as a main constituent of cement can have various effects on consistency depending on its particle size distribution [29,30], it can improve the consistency. Limestone is usually characterized by small particle size distribution, and thus can act as a bearing for bigger particles of cement [31]. On the other hand, limestone has a physical effect on hydration – its particles can act as a nucleation seeds increasing the hydration rate [32], and can prevent a conglomeration of clinker particles, allowing for higher hydration rate [33]. Moreover, in the presence of limestone, during the hydration of CSA clinker, less monosulfate is formed, which leads to a stabilization of ettringite, resulting in higher amounts of etrringite in hydration products [34]. The higher rate of hydration could be responsible for a slight increase in torque of mortars with an addition of 20% and 30% limestone. In case of CSA cement with an addition of 10% of limestone, the plasticizing effect of limestone might be more prevalent.

In table 3 are presented the results of yield stress and plastic viscosity testing of mortars.

Table 3: Yield stress and plastic viscosity of mortars with CSA cement and CEM I 42.5R NA or limestone MW addition.

Rheological property	CSA cement	CEM I 42.5R NA	90% CSA + 10% CEM I	80% CSA + 20% CEM I	70% CSA + 30% CEM I	90% CSA + 10% MW	80% CSA + 20% MW	70% CSA + 30% MW
Yield stress [Nmm]	7.35	15.36	36.4	n/a	n/a	5.51	9.34	8.84
Plastic Viscosity [Nmms]	6.35	7.8	8.66	n/a	n/a	6.35	7.3	6.5

The yield stress conforms to the same relationships as in the case of continuous torque measurement. Due to a high w/c ratio, viscosity of the mortars is consistently low, and differs very little amongst tested mortars. The values could not be calculated for CSA cements with addition of 20% and 30% of CEM I 42.5R NA, due to the extremely fast setting and workability loss, which did not allow to obtain the torque-rotational speed relationship, which would allow to setermine yield stress and plastic viscosity.

3.2 Early shrinkage

The results of early shrinkage measurement are shown in Figures 5 and 6.

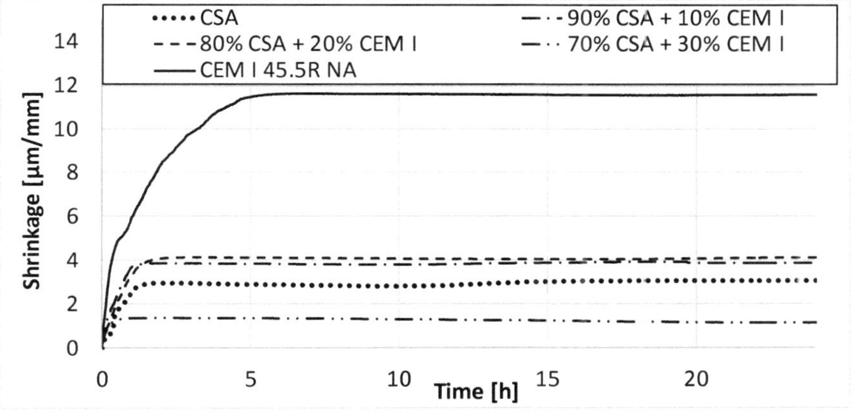

Figure 5: Shrinkage measurement of mortars with CSA and CEM I 42.5R NA in Shrinkage Cone deltaEL in the first 24 h from mixing.

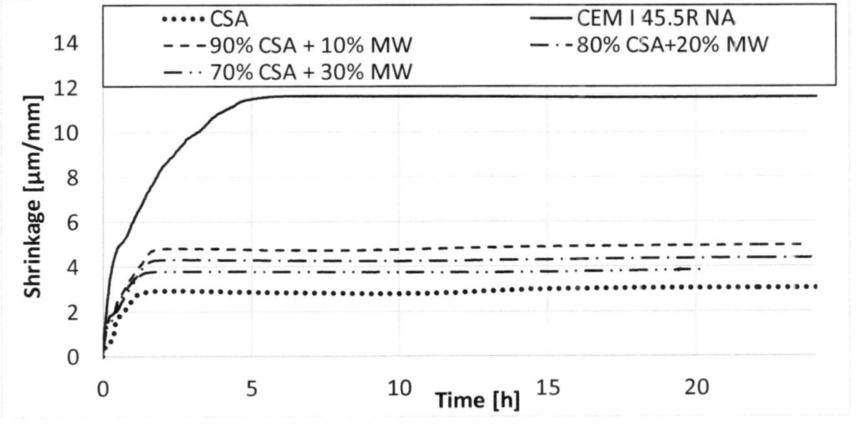

Figure 6: Shrinkage measurement of mortars with CSA mixed with limestone MW in Shrinkage Cone deltaEL in the first 24 h from mixing.

Comparison of early shrinkage of CSA cement and CEM I 42.5R cement show, that the CSA cement was characterized by much lower early shrinkage than in case of Ordinary Portland Cement (OPC) used in the research. Shrinkage of CSA cement was nearly four times lower than in case of CEM I 42.5R NA. It should be also noted, that the time of stabilization of shrinkage is much shorter in case of CSA cement. Addition of 10% and 20% of cement CEM I 42.5R NA to CSA cement resulted in slight increase in shrinkage, while the addition of 30% of CEM I 42.5R NA to CSA cement decreased the shrinkage in comparison even to CSA cement.

Observed effects are connected to hydration process of CSA cement. In the reaction of ye'elemite with water, ettringite forms. The ettringite that forms in this reaction is only partially expansive [5]. Therefore, while chemical shrinkage of CSA cement is considered to be higher than in the case of Portland cement CEM I 42.5R NA [35], the expansive nature of ettringite keeps the early shrinkage low. It should be noted, that the addition of CEM I 42.5R NA in 30% of CSA mass is further reduces the shrinkage. It might be connected to the appearance of expansive ettringite in the hydration of ye'elemite in presence of alumina phases supplied by Portland clinker [36]. In case of lower CEM I 42.5R NA content in the binder, the supply of alumina is lower, and thus the shrinkage is higher.

In case of CSA cements with limestone MW content in range of 10-30%, their shrinkage is slightly higher than in the case of CSA cement, however still significantly lower than in case of Portland cement CEM I 42.5R NA. It could be attributed to the fact that limestone particles, acting as a filler, reduce the capillary pore diameter. This can lead to a slightly increased shrinkage in limestone binders, as shrinkage is directly proportional to the capillary forces which increase as capillary pores decrease in size [30].

4 Results and discussion

The research into the rheological properties of fresh mortars with CSA cement with CEM I 42.5R NA or limestone addition were conducted. Measured was early shrinkage, yield stress and plastic viscosity, and the changes in torque during first 1,5h of hydration. Conducted tests lead to following conclusions:

- Substitution of 10-30% of CSA cement mass with Portland cement CEM I 42.5R NA led to a substantial increase in yield stress and torque, of mortars, and rapid loss of consistency. This effect may occur due to the so-called 'flash setting' of Portland cement in the presence of CSA cement. This topic requires, however, further research.

- Substitution of CSA cement with limestone did not exhibit significant influence on the yield stress or torque, and no rapid loss of consistency was observed.

- Addition of 20% and 30% of limestone led to a small increase. This may be connected with filler effect, in which limestone works increases the rate of hydration, and stabilizing effect of limestone on ettringite formation, which could be more prevalent in case of higher limestone content.

- Addition of 10% of limestone slightly decreased the torque and yield stress, what may be due to the filler effect, which leads to limestone particles acting as a bearing for clinker and sand particles.

- CSA cements were characterized by significantly lower shrinkage than Portland cement CEM I 42.5R NA. Shrinkage of CSA cements with limestone are slightly higher than of CSA cement. This might be due to the higher capillary forces, due to the filler effect of limestone. In case of CSA cement with 30% substitution with Portland cement, a further decrease in shrinkage was observed. This might be due to the expansive forms of ettringite that may appear during hydration of CSA cement in the presence of Portland clinker.

The research indicate, that the substitution of CSA cement with limestone in amount of 10-30 % of cement mass to does not have negative impact on early rheological properties of mortars. Addition of Portland cement CEM I 42.5R NA to CSA cement, in turn, yielded negative effects in terms of yield stress and consistency, while no negative effect was observed for early shrinkage. Further research is required in this topic.

Bibliography

[1] Sharp, J.H., Lawrence, C.D., Yang, R., Calcium sulfoaluminate cements—low-energy cements, special cements or what?, Adv. Cem. Res. 11, 3–131999. doi:10.1680/adcr.1999.11.1.3.

[2] Al Horr, Y., Elhoweris, A., Elsarrag, E., The development of a novel process for the production of calcium sulfoaluminate, Int. J. Sustain. Built Environ. 2017. doi:10.1016/j.ijsbe.2017.12.009.

[3] Imbabi, M.S., Carrigan, C., McKenna, S., Trends and developments in green cement and concrete technology, Int. J. Sustain. Built Environ. 1 194–216, 2012. doi:10.1016/J.IJSBE.2013.05.001.

[4] Miller, S.A., Horvath, A., Monteiro, P.J.M., Readily implementable techniques can cut annual CO2 emissions from the production of concrete by over 20%, Environ. Res. Lett. 11, 2016. doi:10.1088/1748-9326/11/7/074029.

[5] Odler, I., Special Inorganic Cements, E & FN Spon, London, 2000. doi:10.1016/s0008-8846(00)00369-0.

[6] Gartner, E., Barcelo, L., Recherche, L.C. De, The “sustainability” of cement and concrete, Encontro L, 1–45, 2012

[7] Li, G.S., Walenta, G., Gartner, E.M., Formation and Hydration of Low-CO 2 Cements Based on Belite, Calcium Sulfoaluminate and Calcium Aluminoferrite, w: 12th Int. Congr. Chem. Cem., Montreal, Canada, 2007.

[8] Gartner, E., Are There Any Practical Alternatives to the Manufacture of Portland Cement Clinker?, J. CHINESE Ceram. Soc. 40, 61–68 2012.

[9] Julphunthong, P., Joyklad, P., Utilization of several industrialwastes as raw material for calcium sulfoaluminate cement, Materials (Basel). 12, 1–12, 2019. doi:10.3390/ma12203319.

[10] García-Maté, M., De La Torre, A.G., León-Reina, L., Aranda, M.A.G., Santacruz, I., Hydration studies of calcium sulfoaluminate cements blended with fly ash, Cem. Concr. Res., 2013.

[11] Nguyen, K.-S., Nguyen-Ngoc, T.H., Nguyen-Phung, A.T., Do, Q.M., Preparation of calcium sulfoaluminate cement from bauxite/red mud of Tan Rai - Lam Dong, w: 7th Int. Conf. Asian Concr. Fed. (ACF 2016), Hanoi, Vietnam, 2016.

[12] Chen, I.A., Hargis, C.W., Juenger, M.C.G., Understanding expansion in calcium sulfoaluminate-belite cements, Cem. Concr. Res., 2012.

[13] Madan Mohan Reddy, K., Srimurali, M., Bhaskar, M., Mohan ReddyK, M., Characterization of calcium sulfoaluminate cement, 2014. https://www.researchgate.net/publication/287690934

[14] Zhang, L., Su, M., Wang, Y., Development of the use of sulfo- and ferroaluminate cements in China, Adv. Cem. Res. 11, 15–21, 1999. doi:10.1680/adcr.1999.11.1.15.

[15] Winnefeld, F., Martin, L.H.J., Müller, C.J., Lothenbach, B., Using gypsum to control hydration kinetics of CSA cements, Constr. Build. Mater., 2017. doi:10.1016/j.conbuildmat.2017.07.217.

[16] ACI 223R-10 Guide for the Use of Shrinkage-Compensating Concrete, 2010.

[17] Chaunsali, P., Mondal, P., Influence of Calcium Sulfoaluminate (CSA) Cement Content on Expansion and Hydration Behavior of Various Ordinary Portland Cement-CSA Blends, J. Am. Ceram. Soc. 98, 2617–2624, 2015.

[18] Le Saoût, G., Lothenbach, B., Hori, A., Higuchi, T., Winnefeld, F., Hydration of Portland cement with additions of calcium sulfoaluminates, Cem. Concr. Res., 2013. doi:10.1016/j.cemconres.2012.10.011.

[19] Huang, T., Li, B., Yuan, Q., Shi, Z., Xie, Y., Shi, C., Rheological behavior of Portland clinker-calcium sulphoaluminate clinker-anhydrite ternary blend, Cem. Concr. Compos., 2019.

[20] Pelletier-Chaignat, L., Winnefeld, F., Lothenbach, B., Müller, C.J., Beneficial use of limestone filler with calcium sulphoaluminate cement, Constr. Build. Mater., 2012. doi:10.1016/j.conbuildmat.2011.06.065.

[21] Martin, L.H.J., Winnefeld, F., Müller, C.J., Lothenbach, B., Contribution of limestone to the hydration of calcium sulfoaluminate cement, Cem. Concr. Compos. 62, 204–211, 2015. doi:10.1016/J.CEMCONCOMP.2015.07.005.

[22] Winnefeld, F., Martin, L.H.J., Tschopp, E., Müller, C.J., Lothenbach, B., Influence of pozzolanic materials on the hydration of calcium sulfoaluminate cements, Am. Concr. Institute, ACI Spec. Publ. 2017-Janua 167–178, 2017.

[23] EN 197-1:2012 Cement. Composition, specifications and conformity criteria for common cements,

[24] EN 197-2:2014 Cement. Conformity evaluation,

[25] EN 196-3:2016 Methods of testing cement. Determination of setting times and soundness,

[26] Taylor, H.F.W., Cement Chemistry, 2nd editio, Academic Press, London, 1990.

[27] Hargis, C.W., Telesca, A., Monteiro, P.J.M., Calcium sulfoaluminate (Ye'elimite) hydration in the presence of gypsum, calcite, and vaterite, Cem. Concr. Res., 2014. doi:10.1016/j.cemconres.2014.07.004.

[28] Chaunsali, P., Mondal, P., Physico-chemical interaction between mineral admixtures and OPC-calcium sulfoaluminate (CSA) cements and its influence on early-age expansion, Cem. Concr. Res. , 2016.

[29] Bolte, G., Zemente mit hohem Kalksteingehalt, w: 19. Int. Baustofftagung, Weimar,: ss. 405–415, 2018.

[30] Gołaszewski, J., Cygan, G., Gołaszewska, M., Analysis of the Effect of Various Types of Limestone as a Main Constituent of Cement on the Chosen Properties of Cemenet Pastes and Mortars, Arch. Civ. Eng. 65, 75–86, 2019.

[31] Tennis, P.D., Thomas, M.D.A., Weiss, W.J., State-of-the-Art Report on Use of Limestone in Cements at Levels of up to 15 %, USA, 2011.

[32] Zajac, M., Skocek, J., Adu-Amankwah, S., Black, L., Ben Haha, M., Impact of microstructure on the performance of composite cements: Why higher total porosity can result in higher strength, Cem. Concr. Compos. 90 178–192, 2018. doi:10.1016/J.CEMCONCOMP.2018.03.023.

[33] Benachour, Y., Davy, C.A., Skoczylas, F., Houari, H., Effect of a high calcite filler addition upon microstructural, mechanical, shrinkage and transport properties of a mortar, Cem. Concr. Res. 38, 727–736, 2008.

[34] Martin, L.H.J., Winnefeld, F., Müller, C.J., Lothenbach, B., Contribution of limestone to the hydration of calcium sulfoaluminate cement, Cem. Concr. Compos., 2015). doi:10.1016/j.cemconcomp.2015.07.005.

[35] Lura, P., Winnefeld, F., Klemm, S., Simultaneous measurements of heat of hydration and chemical shrinkage on hardening cement pastes, (b.d.). doi:10.1007/s10973-009-0586-2.

[36] Chaunsali, P., Mondal, P., Influence of Calcium Sulfoaluminate (CSA) Cement Content on Expansion and Hydration Behavior of Various Ordinary Portland Cement-CSA Blends, J. Am. Ceram. Soc. 98, 2617–2624, 2015. doi:10.1111/jace.13645.

Are Alkali Activated binder behaviors intermediate between cement and mineral suspensions?

Teresa Liberto[1], Maurizio Bellotto[2], Agathe Robisson[3]

[1,3] Vienna University of Technology, Faculty of Civil Engineering, Research Area Building Materials, Materials Technology E207-01

[2] Politecnico di Milano, Department of Chemistry, Materials and Chemical Engineering G. Natta, Milano, Italy

Abstract

In this work, we compare rheological oscillation measurements on concentrated suspensions of Ordinary Portland Cement (OPC), alkali activated binder (AAB) and calcite. For the three systems, a range of volume concentrations is defined in which these dense suspensions show an "attractive gel-like behavior" [1-3]. The amplitude of this concentration range depends on the extension length of the interparticle forces, and on their nature. We characterize and compare the elastic domains, dominated by the interaction forces, extracting the fractal dimensions of the flocs formed upon self-assembly of the primary particles. We thus provide a hint on the gel heterogeneities and on the deformation mechanisms [1-3]. This preliminary work aims at being a baseline to better understand the nature of the interactions between cement or AAB particles, leading to differences in the fresh state behaviour and setting times, by analyzing the macroscopic rheological response [4]. This innovative approach, obtained from applying a classic fractal model [1], was successfully used to characterize calcite paste [3,4]. The ultimate goal is to control the interaction forces through the fine-tuning of the solid-liquid equilibria, in order to adjust the properties of the paste and fulfill the specific requirements of the final application.

1 Introduction

Alkali-activated binders have been widely discussed and promoted as an essential component of the current and future toolkit of ‘sustainable cementing binder systems’ [5]. However, when selecting a binder alternative to OPC, it is necessary to take into account several aspects beyond the mere mechanical performances and durability properties. First of all, if the binder is used to produce a ready-mixed concrete, it is advisable to use a “one part” material which sets and hardens upon contact with water only, so that the available mixing and processing technologies can be used [6]. Also, with regards to the life-cycle assessment (LCA), the impact of the activating system needs to be duly included. In the case of sodium hydroxide or sodium silicate activation, the LCA will be negatively and significantly impacted [7]. Finally, the fresh-state properties, the response to the addition of superplasticizers and the cohesivity and pumpability of AAB may widely differ from OPC. These requirements are sometimes conflicting with regards to the choice of the AAB system. In the following study, we selected a well-known and long-time used AAB constituted by ground and granulated blast-furnace slag (GGBS) activated with sodium carbonate (or sodium sulphate) [8]. A number of buildings erected with such a binder in the 1950s show little signs of degradation apart from carbonation and are still having compression strength higher than the design values [9]. This binder also has a low environmental footprint, good resistance to environmental attack and it is aesthetically alike to white cement. It is however less cohesive than OPC and does not respond properly to the addition of superplasticizers [10].

The control of cohesivity and the adsorption of superplasticizer are regulated by the surface charge density, surface potential, the ionic composition and ionic force of the interstitial solution [11]. In this study, we measure and compare the rheological properties of OPC and AAB with a model system, a calcite paste, that has been fully characterized in terms of macroscopic and microscopic behavior [3,4]. We interpret the rheological behavior of these suspensions in the framework of the classical fractal elasticity model proposed by Shih et al. [1], with an effort to relate the observed behaviour to the interparticle forces. Our goal is to develop a tool to interpret the role of the surface-solution equilibria in particle attraction and surface polymer adsorption potential.

2 Materials and methods

2.1 Samples preparation

An Ordinary Portland Cement CEM I 52.5 R powder (from Cementi Rossi, Italy, of picnometric density 3140 $kg \cdot m^{-3}$) was mixed directly in distilled water with w/c ratio varying from 0.5 to 0.3 (corresponding to ϕ =0.39-0.51). The mixing was done with a vortex stirrer (Ultra Turrax TD300 from IKA) for three minutes at increasing rotation speed from 2800 rpm to 5800 rpm.

The alkali activated slurry (AAS) was obtained by dispersing GGBS, (from Ecocem France, with D_{50} = 11μm and density of 2944 $kg \cdot m^{-3}$) in distilled water adding sodium carbonate Na_2CO_3 (reagent grade from Sigma Aldrich, density 2540 $kg \cdot m^{-3}$) as solid activator and calcium hydroxide $Ca(OH)_2$ (ventilated hydrated lime CL90-S from Unicalce, Italy, density 2211 $kg \cdot m^{-3}$) as a setting and hardening accelerator. The binder formulation is: 91.5% of GGBS, 5 % of Na_2CO_3 and 3.5 % of $Ca(OH)_2$ (density 2891 $kg \cdot m^{-3}$). The formulation involves stoichiometric amounts of Na_2CO_3 and $Ca(OH)_2$, to optimize both the early setting and the long-term strength [12]. The binder to water ratio is varied from 0.5 to 0.3 (corresponding to ϕ =0.41-0.53). The mixing process is analogous to the one of CEM I.

As detailed in [3], to obtain calcite dense colloidal suspensions, Socal 31 calcite powder (Imerys, average particle diameter 75 nm, density 2710 kg m^{-3}) is also dispersed in distilled water. The maximum range of volume concentrations reached is ϕ = 5–30%. The mixing process was carried out in the same vortex stirrer for five minutes at an increasing mixing rate with the volume concentration from 2800 to 5800 rpm.

2.4 Rheological measurements

Rheological measurements were performed with a stress controlled rotational rheometer (Anton Paar, MCR 302).

Oscillation tests were made with a plate-plate geometry (Anton Paar, PP50 and PP25) at room temperature (20 ± 0.5°C). Plates were either roughened (AAS and CEM I) or covered by sand paper (calcite), in order to avoid wall-slip together with an optimized gap for each system [3]. In particular, the gap is set to

2 mm for the CEM I and the AAS and 4 mm for the calcite paste. A moisture chamber was also used to avoid evaporation, particularly significant in the case of AAS. The "oscillation protocol" is constituted by: (i) a one-minute pre-shear at imposed shear rate of 10 s^{-1} (or deformation γ = 10 %) in which the sample shear stress history is reset, (ii) a time structuration step at low deformation, in the range of elasticity of each system (γ = 0.0005% and t = 2 min for CEM I and AAS - γ = 0.01 % and t = 5 min for calcite) and frequency 1Hz in which the sample reaches a pseudo-steady state, increasing the elastic storage modulus G' and reaching a plateau after 1-2 minutes and (iii) an amplitude sweep at the imposed frequency of 1Hz with increasing deformation from 0.0001 to 10%. The measurements were performed at different volume concentrations, to explore the variation of the elastic properties. As detailed in [3], we extracted a linear storage modulus, G'_{lin} (i.e. storage modulus in the linear regime at low deformation), and a critical strain, γ_{cr} (i.e. value of deformation at the end of the G' linearity), and we compare the different behavior of the three systems as discussed below.

3 Results

Figure 1 and 2 show the influence of the volume fraction ϕ on the elasticity parameters G'_{lin} and γ_{cr} for the three systems.

Starting with calcite, G'_{lin} constantly rises when ϕ increases, with a slope change around ϕ = 17%. This non monotonic trend is strongly confirmed by the γ_{cr} that shows a two-slope scaling behavior [3].

Both CEM I and AAS systems are shifted at a higher volume concentration range corresponding to a water to binder ratio between 0.5 and 0.3. This shift is mainly due to the different particle size and the nature of the interparticle interaction, as detailed in the discussion section.

Looking at both systems in Figure 1, they seem to have a similar behavior. In Figure 2 instead, CEM I shows a γ_{cr} which increases monotonically, while the AAS presents the same two-slope behavior observed for calcite.

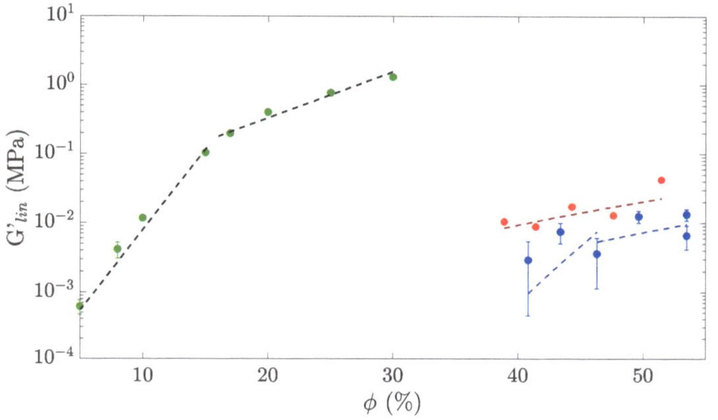

Figure 1: Linear storage modulus G'lin as a function of the volume concentration φ for the three suspensions: calcite (green points), CEM I (red points) and AAS (blue points). The dashed lines represent the plot of the scaling model proposed by Shih et al. [1] as described in the discussion section. Error bars indicate the reproducibility of the results. Notice that the y-axis is in a logaritmic scale.

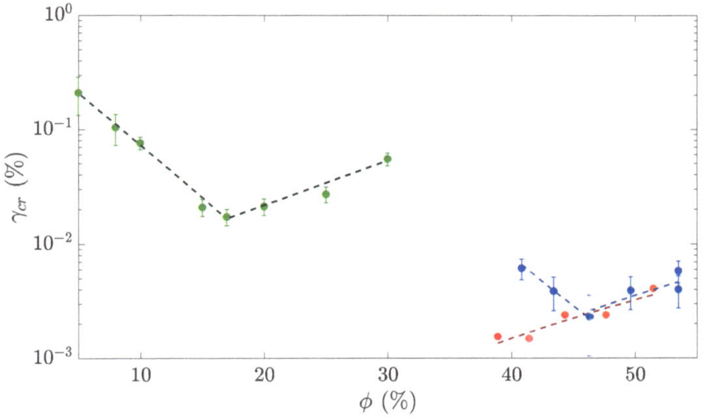

Figure 2: Critical strain γcr as a function of the volume concentration φ for the three suspensions: calcite (green points), CEM I (red points) and AAS (blue points). The dashed lines represent the plot of the scaling model proposed by Shih et al. [1] as described in the discussion section. Error bars indicate the reproducibility of the results. Notice that the y-axis is in a logaritmic scale.

4 Discussion

The amplitude sweeps have been measured, for each system, in the volume concentration range where the system has a gel-like behaviour. The system can flow under shear, with isolated flocs (clusters of particles) dispersed in the continuous interstitial solution, while at rest can form a continuous path of solid particles across the entire volume, which causes the appearance of elastic behaviour with G'>G". We describe the elastic response to the small amplitude deformations according to a classical fractal model proposed by Shih et al. [1]. This model was originally proposed for colloidal suspensions such as our calcite paste, but it can be extended to all systems whose behaviour is controlled by colloidal surface forces, even if the particle dimensions are outside the Brownian motion regime [2].

The structure of the gel network is considered as constituted by fractal flocs (i.e. "repetitive highly porous aggregate composed of smaller primary particles" [13]). The rigidity and the deformation mechanism depend on the density and structure of the flocs, both function of the interaction forces between the particles [14]. Moreover, Shih et al. [1] define two regimes which depend on the compactness of the particle network (i.e. fractal dimension d_f) and which give rise to different trends of G'_{lin} and γ_{cr} as a function of ϕ (Figure 1 and 2). At low ϕ, the floc is porous and deformable under shear. The bond between different flocs is strong compared to the intra-floc bonds, so that the weakest elements are the links between the particles inside the flocs. At high ϕ instead, the flocs are dense and rigid, and the deformation under strain occurs between different flocs. The power law scaling of the elastic modulus and critical strain are: (1) $G'_{lin} \propto \phi^A$, (2) $\gamma_{cr} \propto \phi^B$. Analogously the same equations can be extracted for a critical stress τ_{cr}, resulting from the product of G'_{lin} and γ_{cr}: (3) $\tau_{cr} \propto \phi^C$, where C is the sum of A and B. The scaling exponents are presented in Table 1 directly for the two regimes, respectively deformable and rigid flocs (DF and RF). The detailed derivation of these exponent is explained extensively in [1, 2].

As shown in Table 1, the slope of the exponent is always positive, except for γ_{cr} in the deformable regime. This can be observed in Figure 2 for the calcite and AAS at lower ϕ in the respective volume concentration range. From the scaling laws of G'_{lin}, γ_{cr}, τ_{cr} as a function of ϕ, the fractal dimensions d_f are calculated and reported in Table 2 for the three systems at rest (i.e. small amplitude oscillation measurements). We expected that, under shear (i.e. flow measurements), the continuous, percolated structure formed at rest is destroyed, forming isolated rigid flocs with a resulting floc fractal dimension closer to 3.

Table 1: Exponent of the power law for G'lin and γcr as a function of the fractal dimension df for the different regimes (i.e. DF, Deformable Floc and RF, Rigid Floc).

	DF	**RF**
A ($G'_{lin} \propto \phi^{A}$)	$4/(3-d_f)$	$1/(3-d_f)$
B ($\gamma_{cr} \propto \phi^{B}$)	$-2/(3-d_f)$	$1/(3-d_f)$
C ($\tau_{cr} \propto \phi^{C}$)	$2/(3-d_f)$	$2/(3-d_f)$

Table 2: Fractal dimension df calculated for the three different systems in the two different regimes (i.e. DF, Deformable Floc and RF,Rigid Floc).

	d_f **(DF)**	d_f **(RF)**
CEM I	-	2.71
AAS	2.73	2.75
$CaCO_3$	2.23	2.61

We can notice that for both calcite and AAS suspensions, both regimes are present. As extensively discussed in [2], the presence of both regimes, when exploring the influence of volume fraction only (i.e. no external forces), is not common in colloidal systems and it is surprising to find it also in a non-colloidal one (i.e. AAS). This behavior for our AAS is the sign of weak inter-particle interactions and flocs which become stiffer when growing larger.

Differently, CEM I, despite the similarity with AAS in the high concentrated paste condition state, is presenting only the rigid floc regime throughout the range of ϕ. Before going into detail about the nature of the three different systems, the different deformation mechanisms are schematically illustrated according to the Shih et al. [1] scaling fractal model in Figure 3.

In particular, if we compare Figure 2 with Figure 3, we can recognize the regime of floc deformation (a1 to a3, b1 to b3) and the regime of rigid flocs (a2 to a4, b2 to b4, c1 to c3, c2 to c4).

The different deformation regimes determine microstructural differences of the three systems. The calcite used in this study is a nano-powder which creates, once suspended with water, weak electrostatic attraction between particles. The nature of these interaction (DLVO) was fully described (as detailed in [3]) thanks to the study of the rheology and the physico-chemistry of the suspensions.

Systems like CEM I and AAS are constituted by micrometric particles and, when suspended in water, interact via short-ranged electrostatic and correlation forces that produce short distance electrostatic forces [11]. This explains the narrower volume fraction range at which an CEM I and AAS elastic pastes are formed and which are really sensitive to the complex chemistry of the solution.

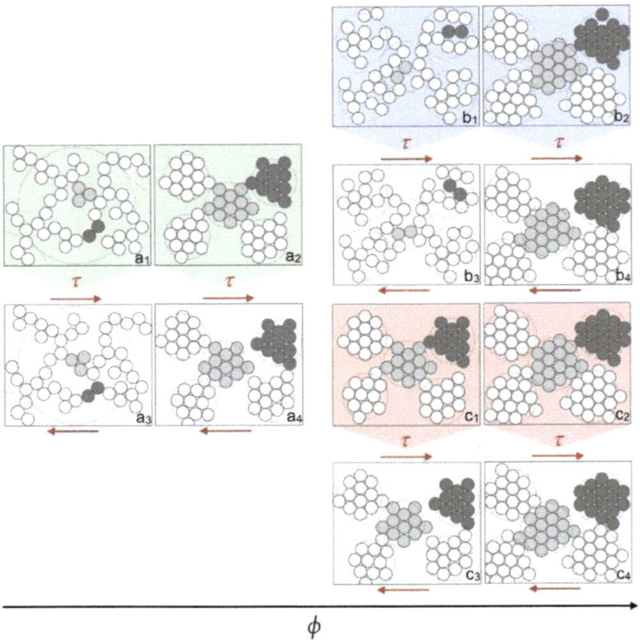

Figure 3: Schematic floc deformation mechanism at different volume concentration φ for the three systems: calcite (a1-a4), AAS (b1-b4) and CEM I (c1-c4). The dashed lines represent the floc. The upper part of each system (light colored) represents flocs at rest and the lower part the new floc configuration after imposing a deformation. Few particles or flocs are colored (i.e. light or dark grey) to help visualize the floc reconfiguration through the movement of the individual particles.

5 Conclusion & Perspective

The observed differences between the rheology of cement and AAS illustrate the different cohesivities of the two pastes. We show on this study that rheology is an important tool to investigate the paste cohesivity, deformation and flow mechanisms, and surface properties. In a follow-up study, we plan to investigate the rheology of the two systems in the presence of superplasticizers, and aim at elucidating the reason behind the high effectiveness of common superplasticizers on cement but not on alkali activated materials.

6 Acknowledgements

We thank Österreichische Bautechnik Vereinigung (ÖBV) and Österreichische Forschungsförderungsgesellschaft (FFG) for their support.

Bibliography

[1] Shih, W.-H., Shih, W. Y., Kim, S.-I., Liu, J. and Aksay, I. A.: Scaling behavior of the elastic properties of colloidal gels. Physical Review A, 1990.

[2] Bellotto, M.: Cement paste prior to setting: A rheological approach. Cement and Concrete Research, 2013

[3] Liberto, T., Le Merrer, M., Barentin, C., Bellotto, M., and Colombani, J.: Elasticity and yielding of a calcite paste: scaling laws in a dense colloidal suspension. Soft Matter, 2017.

[4] Liberto, T., Barentin, C., Bellotto, M., Colombani, J., Costa, A., Gardini, D. and Le Merrer, M: Simple ions control the elasticity of calcite gels via interparticle forces. Journal of Colloid and Interface Science 2019.

[5] Shi, C., Roy, D. and Krivenko, P., Alkali-activated Cements and Concretes. Taylor & Francis, 2006.

[6] Luukkonen, T., Abdollahnejad, Z., Yliniemi, J., Kinnunen, P. and & Illikainen, M.: One-part alkali-activated materials: A review. Cement and Concrete Research, 2017.

[7] Habert, G., d'Espinose de Lacaillerie, J.B. and Roussel, N. : An environmental evaluation of geopolymer based concrete production: reviewing current research trends. Journal of Cleaner Production, 2011

[8] Purdon, A. O.: Improvements in processes of manufacturing cement, mortars and concretes. British Patent GB427227A, 1935.

[9] Buchwald, A., Vanooteghem, M., Gruyaert, E., Hilbig, H., and De Belie N: Purdocement: application of alkali-activated slag cement in Belgium in the 1950s. Materials and Structures, 2015.

[10] Palacios M., Houst Y.F., Bowen P. and Puertas F.: Adsorption of superplasticizer admixtures on alkali-activated slag pastes. Cement and Concrete Results, 2009.

[11] Jönsson, B., Nonat, A., Labbez, C., Cabane, B. and Wennerström, H.: Controlling the Cohesion of Cement Paste. Langmuir, 2005.

[12] Bellotto M., Dalconi M.C., Contessi S., Garbin E. and Artioli G.: Formulation, performance, hydration and rheological behavior of 'just add water' slag-based binders. Proceedings of the 1st International Conference on Innovation in Low-Carbon Cement and Concrete Technology, London 2019.

[13] Jarvis, P., Jefferson, B. and Parsons, S. A.: Measuring Floc Structural Characteristics. Reviews in Environmental Science and Bio/Technology, 2005.

[14] Shih, W.Y., Shih, W.H. and Aksay, I.A.: Elastic and yield behavior of strongly flocculated colloids. Journal of the American Ceramic Society, 1999.

Oszillatorische Rheometrie zur Identifizierung von Strukturabbauprozessen an mineralischen Baustoffen.

Marcel Ramler, Prof. Jürgen Quarg-Vonscheidt
Hochschule Koblenz, Fachbereich Geotechnik

Kurzfassung

Zeitweise fließfähige selbstverdichtende Verfüllbaustoffe (ZFSV) stellen konzentrierte feststoffreiche Suspensionen dar, die vorwiegend zur Verfüllung von Gräben im innerstädtischen Leitungsbau verwendet werden. Aufgrund der Zusammensetzung des ZFSV, welcher tonige Bestandteile und Zement sowie größere inerte Bestandteile wie Sande enthält, bilden sich zeitlich veränderliche Strukturen im Material. Diese Strukturen bedingen maßgeblich die Verformungseigenschaften des Baustoffs. Zudem werden an das Material gegensätzliche Anforderungen gestellt: Zum einen muss das Material fließfähig genug sein um alle Engstellen des Grabens zu füllen, zum anderen muss das Material in jeder Konsistenz in der Lage sein grobe Gesteinskörnung in Schwebe zu halten und gegen Sedimentation stabil zu sein.

In diesem Beitrag soll ein Ansatz zur Diskussion gestellt werden, die sich unter Belastung einstellende strukturelle Veränderung der Suspension zu erfassen und quantitativ zu beschreiben. Dies erfolgt zunächst losgelöst von der klassischen Beschreibung der Fließeigenschaften über die Parameter Fließgrenze $\tau_{f,stat}$ und $\tau_{f,dyn}$ sowie Viskosität μ. Die einsetzende Degeneration von Strukturen soll über die Betrachtung der Arbeit beschreiben werden, welche an der Suspension geleistet werden muss um diese zu verändern oder aufzulösen.

1 Einleitung

Die Eigenschaften von Baustoffen zu verstehen und gezielt zu steuern ist eine wichtige Aufgabe, sowohl aus technologischer als auch ökonomischer und ökologischer Sicht. Nicht ausreichend genau eingestellte Eigenschaften führen dabei unter Umständen zu unerwünschten Erscheinungen oder sogar zu schwerwiegenden Sicherheitsrisiken. Im Falle von mineralischen Baustoffen, welche gleichsam sowohl im flüssigen / fließfähigen Zustand als auch im festen / ausgehärteten Zustand in Erscheinung treten können sind insbesondere die Eigenschaften im fließfähigen Zustand während der Formgebung von großer Bedeutung. Im Falle von Betonen, seien es konventionelle Rüttelbetone oder selbstverdichtende Betone (SVB) sind dies z.B. die Fähigkeit die eingelegte Bewehrung zu umfließen und somit einen störungsfreien Verbundwerkstoff zu bilden. Im Falle von zeitweise fließfähigen und selbstverdichtenden Verfüllbaustoffen [1] (ZFSV oder umgangssprachlich Flüssigboden) gibt es ähnliche Anforderungen hinsichtlich der Verfülleigenschaften, so wird erwartet, dass beispielsweise im Rohrleitungsbau, auch kleine Hohlräume und Zwickelbereiche zwischen oder unter Rohren, durch den Flüssigboden vollständig ausgefüllt werden. Ist das Material nicht imstande die Rohrleitung lückenlos zu umschließen muss damit gerechnet werden, dass die angestrebten optimalen Bettungseigenschaften nicht erreicht werden und es zu Schäden am Rohr kommen kann. Diese zeitweise fließfähigen Baustoffe, zu denen auch Estriche, Mörtel etc. gehören, bestehen aus einem System aus mehreren chemisch oder physikalisch interagierenden Bestandteile welche sich allgemein in eine flüssige und eine feste Phase unterteilen lassen. Somit bilden diese Systeme Suspensionen, deren Eigenschaften sich mit der zunehmenden Bildung fester Strukturen zeitlich verändern.

Zusätzlich zu den Anforderungen an die „Verformungseigenschaften" bzw. die Fließeigenschaften der Suspension im jungen Alter besteht die Erwartung, dass die Suspension in sich stabil bleibt. Werden feste Bestandteile der Suspension z.B. unter Einwirkung der Schwerkraft nicht mehr durch die flüssige Phase am Absinken gehindert erfolgt eine, wenigstens teilweise, Trennung der Phasen, wobei zunächst die festen Bestandteile sedimentieren. Dieses Stabilitätsproblem der Suspension wird in verschiedenen Baustoffen beobachtet (SVB, Rüttelbetone, ZFSV).

Konventionelle Rüttelbetone verlieren u.U. durch die dynamische Belastung während der Verdichtung ihre tragende Struktur, CSH-Phasen innerhalb der Zementleimmatrix werden aufgetrennt und Agglomerate aufgelöst. Selbstverdichtende Betone und ZFSV definieren sich über einen zunächst relativ geringen Fließwiderstand, so dass diese u.U. wesentlich empfindlicher gegenüber Sedimentationserscheinungen sind.

Untersuchungen zur Entwicklung und zum Verhalten von Strukturen unter Belastung sollen zu einem besseren Verständnis der Mechanismen und Abläufe in feststoffreichen Suspensionen beitragen um schließlich die Eigenschaften gezielt beeinflussen zu können. Um die Veränderung von Strukturen zu erfassen wurden Versuche mittels modifizierter ViskoWaage, welche miniaturisierte Pfahlzugversuche darstellen, sowie oszillatorische Versuche mittels Viskomat NT der Firma Schleibinger durchgeführt.

2 Strukturwiderstand

Nachfolgend soll ein kurzer Überblick darüber gegeben werden, was als maßgebliche Struktur einer Suspension angesehen werden kann und wie sich die ausgebildete Struktur bei Einwirkung von Kraft oder bei dynamischer Anregung verhält. Speziell die Veränderung und der Abbau / Umbau von Strukturen hat wesentlichen Einfluss auf die Fließeigenschaften von Suspensionen.

Beispielsweise können Strukturen wie bei Betonen aus sich bildenden CSH-Phasen innerhalb des Zementleims mit zusätzlichen Anteilen aus makroskopischen Korn-zu-Korn-Kontakten oder aber wie im Bereich der fließfähigen Verfüllbaustoffen aus der typischen Kartenhausstruktur der Schichtsilikate [6] mit kleineren Anteile der CSH-Phasen infolge des zugegebenen Zements bestehen. Dabei soll beachtet werden, dass die Bildung von CSH-Phasen in ZFSV nur von untergeordneter Rolle sein soll, da der Zement nur zur Kontrolle des freien Wassers der Suspension zugegeben wird. Eine zu starke Ausbildung der CSH-Phasen, ähnlich konventioneller Betone führt zu einer stark eingeschränkten Wiederaushubfähigkeit.

3 Energetische Betrachtung von Fließzuständen

An dieser Stelle wird kurz die energetische Betrachtung welche dem Parameter der Strukturabbauarbeit erläutert. Der Begriff der Strukturabbauarbeit soll die Energie beschreiben, die einer Suspension bzw. der Struktur einer Suspension in Form von mechanischer Arbeit zugeführt werden muss um eben diese Strukturen überwinden. Durch Aufbringung von externer Kraft (Volumenkräfte infolge der Erdbeschleunigung oder Oberflächenkräfte infolge Auflasten etc.) verformt sich ein finites Fluidelement, innere Rückhaltekräfte (als Analogie ist eine gespannte Feder vorstellbar) sorgen für den noch erreichbaren Gleichgewichtszustand bzw. eine Aufrechterhaltung der Struktur.

Wird die aufgebrachte Kraft gesteigert so werden die inneren Haltekräfte im Fluidelement ab einem zu bestimmenden Grenzwert überschritten, die Struktur bricht. Während diesem Prozess leisten die inneren Kräfte im Fluidelement Arbeit. Die Arbeit welche an dem Element, bis zum Strukturbruch geleistet werden muss soll die Strukturabbauarbeit sein. Der im Versuch mittels ViskoWaage bestimmte Ausziehwiderstand eines Verankerungskörpers aus einer konzentrierten Suspension zeigt einen charakteristischen Verlauf, wie in Abbildung 1 dargestellt. Im Bereich des ersten Kraft-Peaks bis zum Erreichen des ersten Minimalwertes zeigt sich ein markanter nichtlinearer Weg-Zeit-Verlauf.

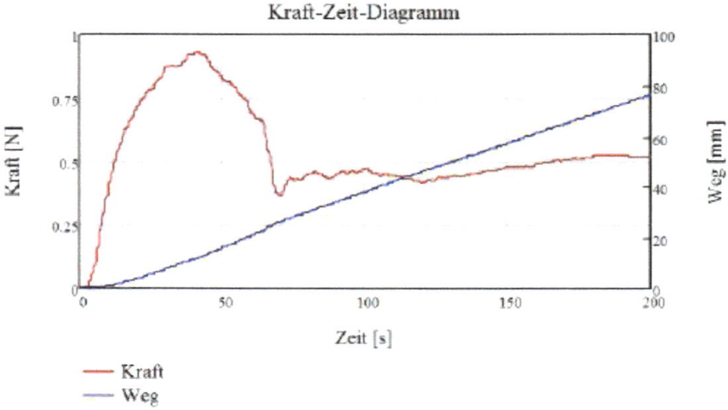

Abbildung 1: Charakteristischer Messwertverlauf.

Innerhalb dieses Bereiches werden Widerstände der haltenden Strukturen, welche einer Bewegung entgegenstehen mit zunehmender Verformung mobilisiert und bei Erreichen der Grenzkraft zerstört. Die Folge sind abfallende haltende Kräfte und ein Anstieg der möglichen sich einstellenden Verformungsgeschwindigkeit. Mit Erreichen der minimalen Widerstandskraft stellt sich ein verformungsunabhängiger Restwert des Strukturwiderstands ein. Der anschließend zurückgelegte Weg wird mit einer konstanten Geschwindigkeit zurückgelegt.

Durch Auswertung der erfassten Daten (Kraft, Weg und Zeit) welche mehrmals in der Sekunde aufgezeichnet werden, ist es möglich das numerische Integral der Kraft über den Weg über mehrere Stützstellen zu bilden. Integriert man somit die gemessene Kraft über den zurückgelegten weg (welcher innerhalb der noch intakten Struktur ein zeitliches Differential $\frac{d^2s}{dt^2} = \ddot{s} \neq 0$ auf eine Beschleunigung zurückzuführen ist). Führt man das die Integration bis zum Übergangspunkt zwischen $\frac{d^2s}{dt^2} \neq 0$ und $\frac{d^2s}{dt^2} = 0$ im Gleichgewicht des Bewegungszustandes s_{eq} aus, so kann die verrichtete Arbeit numerisch ausgewertet werden (vgl. Abbildung 2).

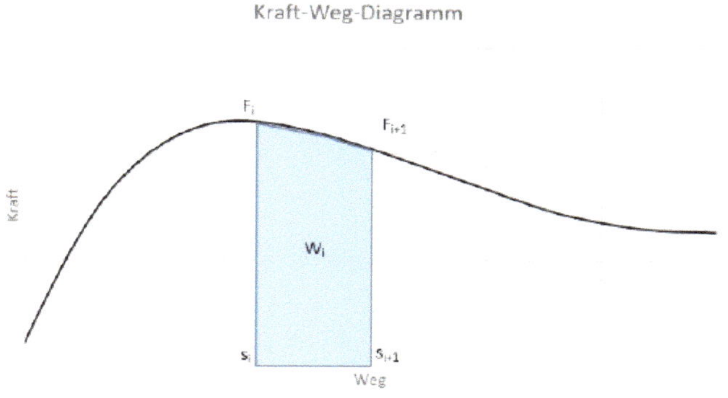

Abbildung 2: Schematischer Kraft-Weg-Verlauf einer charakteristischen Suspension.

$$W = \int_{s_0}^{s_{eq}} F(s)ds = \frac{1}{2}\sum_{s_0}^{s_{eq}} [(F_i + F_{i+1}) * (s_{i+1} - s_i)] \tag{1}$$

Anhand eines Beispiels soll der Nutzen der Betrachtung der Strukturabbauarbeit gezeigt werden. Abbildung 3 zeigt den Sedimentationsgrad des grobkörnigen Zuschlagmaterials einer fließfähigen SVB-ähnlich Suspension unter dynamischer Einwirkung abhängig von der gemessenen Fließgrenze der Suspension. Die Mischungen wurden dabei unter 50 und 75 Hz dynamisch angeregt. Diese Sedimentation, also das Absinken der Partikel und die daraus folgende Entmischung, wurden über die zuvor gemessene dynamische Fließgrenze [5] aufgezeichnet. Man erkennt die Tendenz einer steigenden Sedimentation bei kleineren dynamischen Fließgrenzen, jedoch ist die Bestimmtheit dieser Beobachtung nur schwach ausgeprägt.

In Abbildung 4 ist stattdessen die aufgetretene Sedimentation (in Anlehnung an [2]) unter dynamischer Einwirkung auf die notwendige Strukturabbauarbeit bezogen. Man erkennt deutlich, dass bei einer größeren notwendigen Arbeit, um die Strukturen zu zerstören oder zu überwinden wesentlich geringere Sedimentationserscheinungen beobachtet werden können. Dies lässt sich damit begründen, dass die aufgebrachte dynamische Anregung in Teilen, welche von der Strukturstärke / Strukturabbauarbeit abhängen, von der tragenden Suspension aufgenommen werden können. Erst wenn die erforderliche Strukturabbauarbeit im Fluid geleistet wurde, werden Strukturen in solch einem Ausmaß zerstört, dass das Grobkornmaterial nicht mehr in seiner relativen Position gehalten werden kann. Je kleiner die dafür notwendige Arbeit, desto größer bilden sich Sedimentationseffekte aus. Es ist zudem erkennbar, dass sich das Verhalten bzw. die Ausgleichsfunktion bei zunehmender dynamischer Belastung parallel verschiebt.

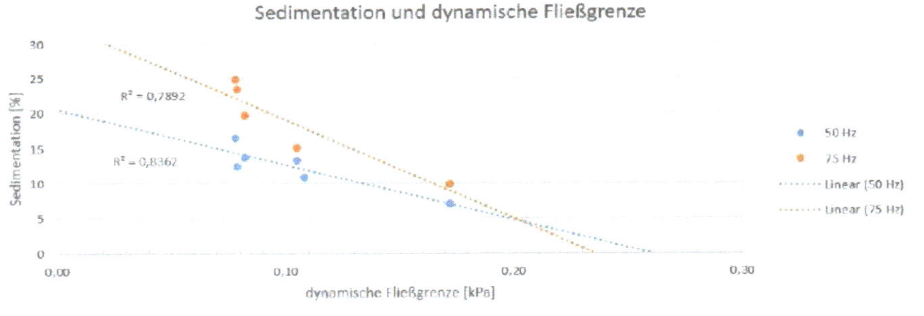

Abbildung 3: Sedimentation bezogen auf die dynamische Fließgrenze.

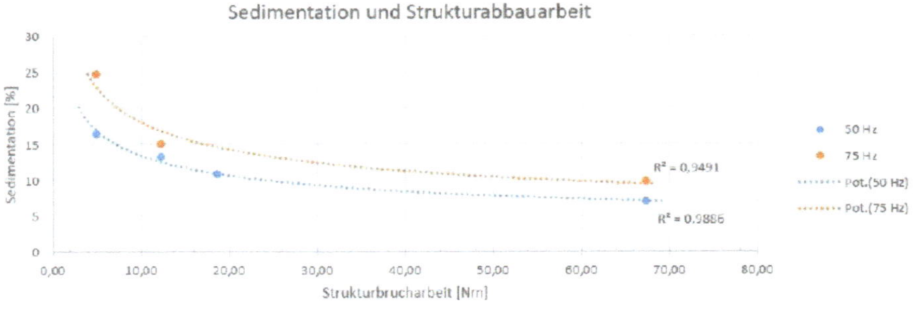

Abbildung 4: Sedimentation bezogen auf die notwendige Strukturabbauarbeit.

3 Oszillatorische Grundlagen

Mittels Viskomat NT soll untersucht werden, wie durch Krafteinwirkung auf ein idealisiertes Partikel in Kugelgestalt der Bewegungswiderstand infolge der Struktur der umgebenden Suspension verändert wird. Als Probengeometrie dient eine Stahlkugel mit Durchmesser 20 mm. Um den Strukturabbau zu erfassen wurden oszillatorische Messprogramme gefahren.

Speziell die Phasenverschiebung der registrierten Kraft am Probekörper bezogen auf die einwirkende Auslenkung des Probentopfes gibt Ausschluss über etwaige Veränderungen der Suspensionssturktur. Verläuft die Kraftantwort in Phase der aufgebrachten Auslenkung so lässt dieses idealelastische Verhalten auf eine ungestörte Struktur schließen. Die inneren Haltekräfte bleiben intakt. Entsteht, bei größer werdenden Belastung (Amplitude oder Frequenz) eine Phasenverschiebung, so kann im Umkehrschluss auf Strukturabbau geschlossen werden (Abbildung 5). Der Phasenwinkel kann Werte zwischen $0° \leq \delta \leq 90°$ annehmen. Wobei $\delta = 0°$ auf ideal elastisches Materialverhalten unterhalb der Fließgrenze hindeutet und $\delta = 90°$ auf ideal viskoses Verhalten bedeutet in dem alle haltenden Strukturen zerbrochen oder überwunden sind. Zusammengefasst lassen sich beide Anteile der Spannung wie folgt darstellen [3]:

$$\sigma = G'\hat{\gamma}\sin(\omega t) + G''\hat{\gamma}\cos(\omega t)$$

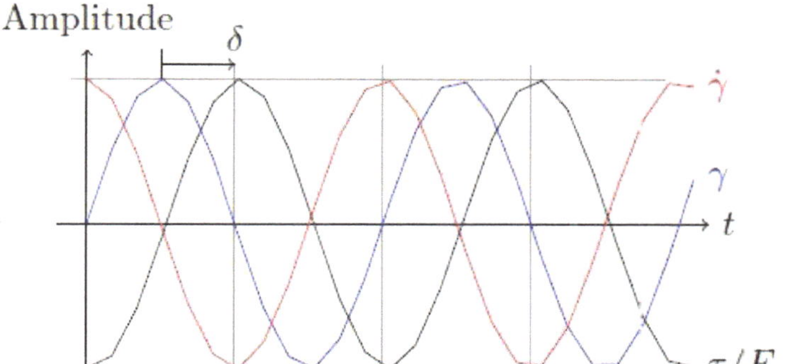

Abbildung 5: Oszillationsverhalten.

4 ViskoWaage

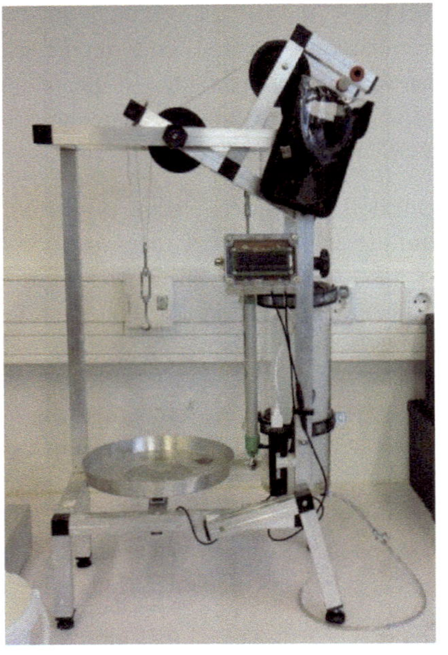

Eine robuste Variante die Fließgrenze zu ermitteln ist die modifizierte Visko-Waage [4]. In Anlehnung an den Versuch mit der ursprünglichen ViskoWaage wird kraftgesteuert Belastung auf die Suspension aufgebracht und der Punkt, an welchem die Suspension der Spannung nicht mehr standhalten kann.

Bei der modifizierten ViskoWaage wird ein Verankerungskörper in die Suspension eingedreht, dies führt durch die Form des Verankerungskörpers zu minimalen Strukturstörungen. Bezieht man die registrierten Kräfte an der Suspension auf die momentan benetzte Fläche des Verankerungskörpers lassen sich die einwirkenden Schubspannungen auf die Suspension berechnen. Die Fließgrenze ermittelt sich dabei wie folgt:

$$\tau_F = \frac{(G_{max} + \rho_{ZFSV} V_{Dübel} - G_{Rest}) * g}{A_{Dübel}} \qquad (2)$$

5 Versuchsprogramm

Die ersten Materialien, welche mit dem Fokus auf die notwendige Strukturabbauarbeit untersucht wurden, waren reine Bentonitsuspensionen (ein Grundbestandteil der ZFSV) verwendet. Auf Zugabe von weiteren Inhaltsstoffen (Kornmaterial oder Zement) wurde verzichtet um die einzelnen sich bildenden Strukturen getrennt voneinander beobachten zu können.

5.1 Untersuchtes Material

Es wurden Tonsuspensionen in unterschiedlichen Konzentrationen hergestellt und sowohl mittels ViskoWaage als auch mittels Viskomat NT untersucht. Um auch die zeitliche Entwicklung der für Bentonit typischen Kartenhausstruktur erfassen zu können wurden die Suspensionen nach Herstellung zunächst intensiv durchmischt, dies stört die Kartenhausstuktur. Die weiteren Versuche erfolgten nach definierten Ruhephasen. Folgendes Messregime wurde dabei angewandt:

1. Bentonit-Wasser: 6% Konzentration: 0, 5, 10, 30, 60 Min. nach Herstellung
2. Bentonit-Wasser: 8% Konzentration: 0, 5, 10, 30, 60 Min. nach Herstellung
3. Bentonit-Wasser: 10% Konzentration; 0, 5, 10, 30, 60 Min. nach Herstellung

5.2 Versuchsreihe 1: ViskoWaage

Mittels ViskoWaage wurden die Parameter der statischen Fließgrenze und der aufgebrachten Strukturabbauarbeit nach Gleichung 1 ermittelt. In Abbildung 6 ist zu erkennbar, dass sich die Fließgrenze abhängig von der Zeitdauer nach Durchmischung ändert. Dabei ist die Entwicklung der Fließgrenze an Mischung 1 (Bentonit-Wasser-Suspension mit 6 % Konzentration) kaum auszumachen. Entsprechend zeigt sich bei Mischung 3 (10% Konzentration) ein deutlicher Anstieg der Fließgrenze. Korrespondierend dazu, wenn auch stärker schwankend, bei zunehmender Fließgrenze auch ein Anstieg der aufzubringenden Strukturabbauarbeit. So zeigt sich bei Mischung 3 ein deutlicher Anstieg der aufzubringenden Arbeit bis zum Strukturbruch, wogegen bei Mischung 2 und 1 der Anstieg weniger stark verläuft. Auch ist erkennbar, dass die Konzentration des Bentonits einen starken Einfluss auf den sich entwickelnden Strukturwiderstand hat.

5.3 Versuchsreihe 2: Konstante Amplitude und Frequenz

In dieser Versuchsreihe wurde das Material konstant bei 0,5° Winkelverdrehung des Versuchsbehälters und einer Frequenz von 0,1 Hz bewegt. Innerhalb dieser Versuchsreihe zeigten sich keine durch eine Phasenverschiebung gekennzeichnete Bruchzustände.

In Abbildung 7 ist der Verlauf eines exemplarischen Versuchs dargestellt. Zu erkennen ist der quasi lineare Zusammenhang zwischen Spannung und Auslenkung. Etwaige Öffnungen der Hystereseschleife lassen sich eher an versuchstechnischen Ungenauigkeiten als an wirklicher Phasenverschiebung festmachen.

Neben der Beobachtung, dass das Material nicht in einen viskoelastischen Zustand übergeht, konnte festgestellt werden, dass das Material sehr wohl einen Fließwiderstand mit der Zeit aufbaut. In Tabelle 1 dargestellt sind die Widerstandsmomente, welche am Viskomat NT registriert wurden.

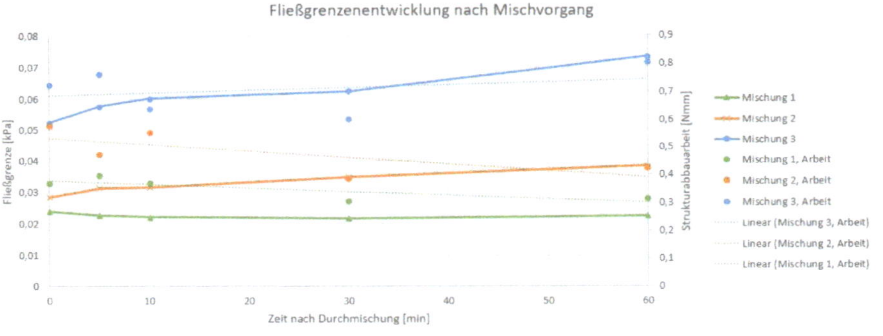

Abbildung 6: Entwicklung der Fließgrenze und der Struktur.

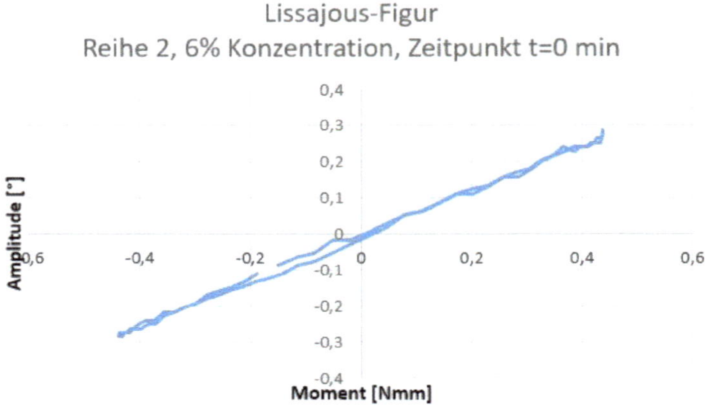

Abbildung 7: Exemplarischer Verlauf im Phasenraum.

Tabelle 1: Widerstandsmomente an Reihe 1

Zeit nach Durchmischen [min]	max M [Nmm]		
	Material 1 6%	Material 2 8%	Material 3 10%
0	0,219	0,682	1,804
5	0,443	0,898	2,051
10	0,466	0,828	3,720
30	0,554	1,292	4,306
60	0,423	1,404	2,831

5.4 Versuchsreihe 3: Ansteigende Amplitude, feste Frequenz

In dieser Versuchsreihe wurde der Auslenkwinkel stufenweise von 0,1° bis 1° gesteigert, die Frequenz blieb konstant bei 0,1 Hz.

Die Steigerung des Auslenkwinkels bei gleichbleibender Frequenz hat eine größere Relativgeschwindigkeit zwischen Suspension und Probenkörper zur Folge. So wurden bei dieser Versuchsreihe Geschwindigkeiten von 0,0330 mm/s linear steigend bis 0,3299 erreicht, was einer Verzehnfachung der Relativgeschwindigkeit zwischen Kugel und Suspension entspricht.

Auffällig in dieser Reihe war besonders, dass bei beginnendem Strukturbruch neben der beschriebenen Phasenverschiebung zusätzlich auch die Amplitude der mobilisierbaren Widerstände abnimmt. Exemplarisch ist dies in Abbildung 8 anhand Mischung 1 (6 % Bentonitsuspension) dargestellt. In der Lissajous-Figur Abbildung 9 zeigt sich zudem das „Aufweiten" der Hystereseschleife, welche auf Phasenverschiebung hindeutet.

Dieser Effekt zeigte sich nach längerer Standzeit des Materials (von 0 min bis 60 min nach Durchmischung) bei verschiedenen Mischungen und unterschiedlichen Belastungen.

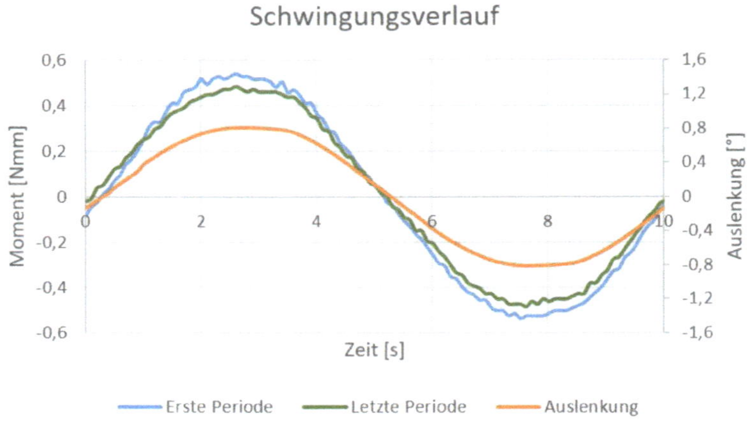

Abbildung 8: Exemplarischer Schwingungsverlauf am Beispiel der Mischung 1.

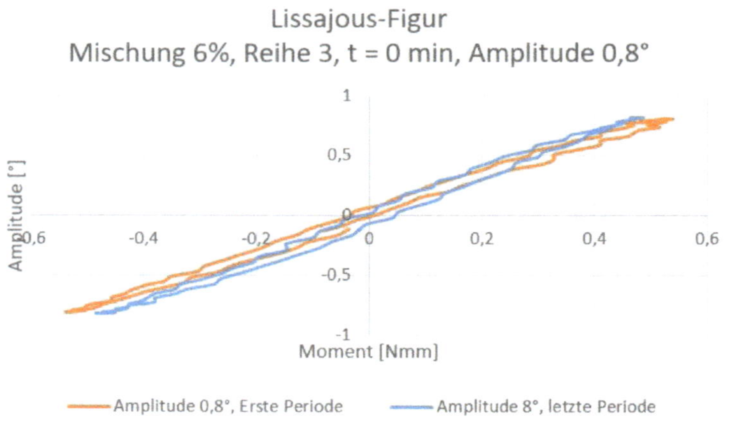

Abbildung 9: Lissajous-Figur, Versuchsreihe 3, 6% Mischung, 0 Minuten nach Mischung bei Amplitude 0,8°.

Ein Nachteil dieser Versuchsreihe war, dass bei gleichzeitigem Steigern der Amplitude auch die Geschwindigkeit bei gleichbleibender Frequenz am Prüfkörper linear zunimmt. Eine Trennung von Effekten infolge Auslenkung und Effekten infolge Geschwindigkeit war nicht möglich.

5.5 Versuchsreihe 4: Konstante Amplitude, steigende Frequenz

In dieser Reihe wurde das Verhalten des Materials isoliert in Abhängigkeit der Geschwindigkeit betrachtet. In den durchgeführten Versuchen stieg die Geschwindigkeit linear von 0,1640 mm/s bei einer Frequenz von 0,1 Hz mit einer Amplitude von 0,5° bis zu einer Geschwindigkeit von 1,640 mm/s bei einer Frequenz von 1,0 Hz bei gleichbleibender Amplitude von 0,5°. Die Amplitude, so haben die vorangegangen Versuche gezeigt, zeigte mit gleichbleibender Frequenz von 0,1 Hz keine Phasenverschiebung und somit keinen Bruch der Strukturen.

Eindeutig erkennbar ist, dass der Veformungswiderstand des Materials nicht ausschließlich von der Auslenkung abhängt, so zeigt sich in Abbildung 10 eine deutliche Weitung in der Lissajous-Figur welche auf eine Phasenverschiebung hindeutet. Anders als bei Reihe 3 liegen die Werte des Widerstandsmomentes innerhalb einer Frequenzspanne übereinander. In Abbildung 9 hingegen sieht man, dass der Widerstand innerhalb der Amplitudenspanne weiter abbaut.

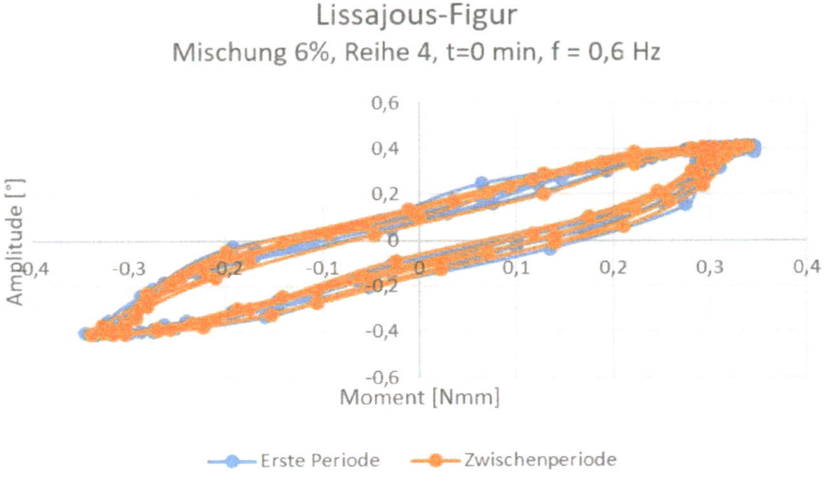

Abbildung 10: Lissajous-Figur, Reihe 3, 6% Mischung.

6 Fazit

Im Rahmen von Untersuchungen zur Suspensionsstabilität verschiedener Rüttelbetone konnte eine Abhängigkeit der Sedimentationsneigung von den Fließeigenschaften der untersuchten Materialien nachgewiesen werden. Als Parameter zur Beschreibung der Fließeigenschaften wurden zunächst die statische sowie die dynamische Fließgrenze der hochkonzentrierten Suspensionen mittels modifizierter ViskoWaage gemessen. Eine wesentliche Verbesserung der Korrelation zwischen den Messwerten und der beobachteten Suspensionsstabilität konnte durch die Erweiterung der Betrachtung auf die tatsächlich geleistete Arbeit zur Überwindung des Herausziehwiderstandes der Suspension gegenüber den eingeschraubten Verankerungskörpern der ViskoWaage erreicht werden. Da zum Herausziehen der Einschraubkörper die Struktur des untersuchten Materials überwunden werden muss, so wie die Struktur überwunden wird bevor ein Zuschlagkorn aus der Suspension absinken kann, wird die erforderliche Strukturabbauarbeit als Maß zur Beurteilung der Suspensionsstabilität identifiziert.

Die Idealisierung der Bewegung eines Gesteinskorns während der Sedimentation stellt die Untersuchung der Kugelumströmung im Viskomat NT dar. In entsprechenden Versuchen an Bentonitsuspensionen wurde die notwendige Arbeit zur Überwindung der Suspensionsstruktur als Ursache für den Bewegungswiderstand gegenüber der Kugel gemessen. Eine Korrelation der Strukturabbauarbeit in den Versuchen an der ViskoWaage und den Versuchen am Viskomat NT bestätigt die Beobachtung, dass die Sedimentationsneigung einer Suspension über die Struktur-abbauarbeit beschrieben werden kann (vgl. Abbildung 11).

Die Beurteilung der Sedimentationsstabilität gelingt bislang nur im Rahmen von Korrelationen zwischen gemessener Strukturabbauarbeit und beobachteter Sedimentation. Weitere Untersuchungen sollen demnächst eine eindeutige Definition des Phänomens der Struktur sowie die Identifizierung der maßgeblichen Para-meter zur Quantifizierung des Strukturwiderstandes ermöglichen. Auf der Grundlage dieser Untersuchungen wird eine allgemeingültige Berechnung der Sedimentationsstabilität auf der Basis eindeutiger Messwerte möglich sein.

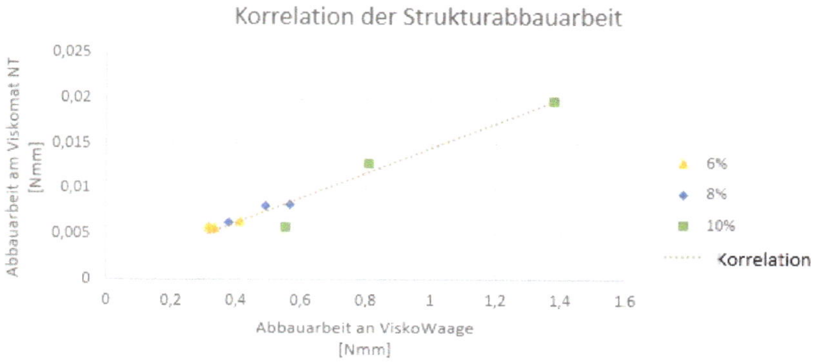

Abbildung 11: Darstellung der Korrelation der Strukturabbauarbeiten.

Literatur

[1] H ZFSV, Forschungsgesellschaft für Straßen- und Verkehrswesen, Arbeitsgruppe „Erd- und Grundbau“: Hinweise für die Herstellung und Verwendung von zeitweise fließfähigen, selbstverdichtenden Verfüllbaustoffen im Erdbau. FGSV Verlag, Köln, 2012

[2] BAW Merkblatt (Entwurf), Entmischungsstabilität von Beton, Karlsruhe, 2018

[3] Das Rheologie Handbuch, Metzger T., Vincentz Network, Hannover 2016

[4] Quarg-Vonscheidt, Sosinka: Ein Vorschlag zur einfachen Bestimmung der Fließgrenze nicht-newtonscher Fluide als physikalischer Parameter zur Charakterisierung fließfähiger Baustoffe. Bericht, Hochschule Koblenz, 2018

[5] Fleischmann, Ein Beitrag zur Bestimmung der rheologischen Eigenschaften selbstverdichtender Betone mit dem Kugel-Messsystem.

[6] Praetorius, Schößer: Bentonithandbuch. Ringspaltschmierung für den Rohrvortrieb. Ernst & Sohn GmbH & Co.KG. 2016

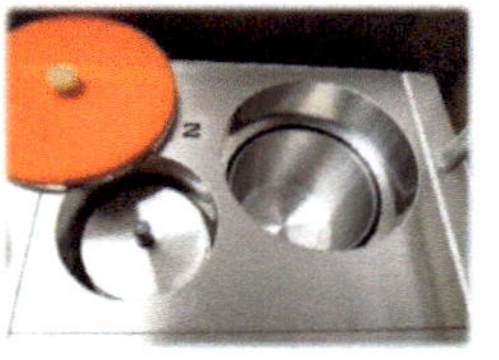

Investigations on maximum fibre content for injecting high-performance mortars

Ludwig Hertwig, Klaus Holschemacher
HTWK Leipzig, Structural Concrete Institute, Leipzig, Germany
Phone: 0341-3076 8832, e-mail: ludwig.hertwig@htwk-leipzig.de

Abstract

Injection is a novel manufacturing process for high-performance mortars. Especially in connection with fibres or textile reinforcement it represents a powerful alternative to the pouring process. Filling from below under pressure ensures a wider use of high-performance mortars regarding rheological properties such as viscosity and yield stress. The procedure follows the desire for slim, lightweight and load-compliant design. A successful injection is in particular depending on fresh mortar properties (viscosity, yield stress, fibre content) and geometrical barriers (formwork, reinforcement). The use of short fibres in mixtures changes the rheological properties. From a static point of view, high fibre contents are desirable, but decreasing workability limits them. The aim of this work is to suggest a fibre content, which improves the static properties and still allows injection. A self-compacting and injectable dry mix mortar (M0) with a maximum grain size Dmax = 2 mm is available. Seven mixtures with different dosages (f = 0.00 vol.-% – 0.66 vol.-%) of two Polyvinyl alcohol (abbreviated PVA)-fibre types with aspect ratios l/d = 55 and 77 were investigated. The use of the Sliding Pipe Rheometer (abbreviated Sliper) provided information on pressure-flow rate relationship and reference values of rheological parameters to estimate. The rheological tests took place at t = 15 min, 30 min and 45 min after initial water addition. The fibres influenced the formation of the lubricating layers and changed the measuring method. The evaluation, together with the results of the hardened concrete tests after t = 7 d, 14 d and 28 d, resulted in a recommendation of a maximum fibre content.

1 Introduction

The development and application of high-performance concretes and mortars[1] follows the desire for slim, lightweight and load-compliant design.

pumping sump
600 x 600 x 800 mm
wall thickness d = 25 mm

segment canoe
700 x 300 mm
wall thickness d = 5 mm

canoe
5200 x 700 x 300 mm
wall thickness d = 5 mm

Figure 1: Prototypes successfully manufactured using injection technology.

The production of complex and thin components by pouring is highly dependent on the rheological properties of the mixture. However, since self-compacting concretes and mortars are susceptible to segregation [2], the formation of the

[1] High performance concrete is widely regarded as concrete with compressive strength of $f_{cm} \geq 60$ MPa [1]. Here, the concept „high performance mortar“ (abbreviated HPM) combines the demands on high compressive strength, small maximum grain size $D_{max} \leq 2$ mm, high demand on workability and usability (architectural concrete).

maximum flowability is risky. In addition, the process is time-consuming and labour-intensive, subjective application errors can easily occur. The injection[2] process meets these challenges and therefore, it is an effective process for the production of high-performance mortars. Figure 1 demonstrates the potential showing successfully injected prototypes. The process uses the compression and distribution energy of the applied pump pressure. This may reduce processing time significantly and lower the binder and superplasticizer (SP) contents. Filling from below ensures that the fresh concrete level in the formwork rises evenly. The fresh concrete level increases constantly and thus the pump presses the matrix from below against flow barriers (reinforcement, geometry). The entire fresh concrete is in motion and thus spreads better.

Mixtures must meet rheological requirements for successful injection. The demands regarding pumpability also apply to injectability. In general, pumpability of concrete depends on the relation between pumping pressure P and the corresponding flow rate Q. The prediction and handling of pumping concrete is described in various scientific publications [4–9]. One way to determine the characteristics of pumping concrete is the Sliding Pipe Rheometer (Sliper). The device simulates the flowing process in an open pipe end [5] and above all measures the properties of the lubricating layer [7]. The mortars are loaded in the pipe lifted-up to start position. Starting the measurement the pipe slides down and shears the lubrication layer around the concrete plug [5]. Repetitions at different load steps simulate various flow rates. Time, distance and pressure data build a rheological model of the lubricating layer. The parameters a and b of Sliper device describe the rheology where a represents yield stress τ_0 and b plastic viscosity μ [9]. With regard to the rheological properties of mixtures with fibres, Ghanbari and Karihaloo [10] proposed a calculation model for plastic viscosity. Based on the plastic viscosity of the paste, the model considers the fibres using mechanical principles. Gerland et al. [11] noticed a linear increase of viscosity with increasing fibre content in silicone oil.

[2] In the field of crack repair, the injection method describes the pressing of suspensions containing binding agents into concrete [3]. Since the pumping of concrete starts from a pipe diameter of $d \geq 80$ mm [3] the term injection is used to describe the analogy of filling complex geometries with HPMs under pressure.

2 Experimental investigations

2.1 Materials and test setup

For the injectability of mortars, there are currently no recommendations regarding the mix design. Therefore, it was assumed within the scope of these investigations that the mixing concept of a self-compacting mortar is target-oriented. A self-compacting dry mix mortar with a maximum grain size of $D_{max} = 2$ mm and a constant workability of $t = 30$ min was available.

Two types of fibres with different dosages added to the basic mixture (Table 1) built the test matrix. The water-binder value thus remained almost constant, but the volume proportions changed. PVA fibres showed good results in various experiments [12] with regard to rheology and load-bearing behaviour and formed the basis for the experiments. In previous studies a pump sump with AR glass fibres (fibre content $f = 0.75$ vol.-%, length $l = 12$ mm) could be produced by injection [13]. Table 2 shows the properties of the fibres used, the difference was mainly the fibre thickness d. F2 was much thinner, the difference in l/d aspect ratio was $\Delta l/d = 22.5$. Figure 2 also shows the optical differences: The fibre geometry of F1 was cylindrical, whereas F2 was more flexible and not straight. With a comparable density and length, the smaller diameter of F2 caused an increase in the number of fibres in the mixtures, as the fibre content remained constant.

Table 1: Mix designs of used mortar showing masses, water-binder-ratio (w/b) and calculated density ρfr.

mix	fibre type	dosage f	water	cement	min. additives	agg.	SP	fibres	w/b	ρ_{fr}[a]
		[vol.-%]			[kg/m³]				[−]	[kg/m³]
M0	-	-	239	567	113	1198	4.0	-	0.38	2120
M1	F1	0.33	239	567	113	1198	4.0	21	0.38	2141
M2	F1	0.66	239	567	113	1198	4.0	42	0.38	2162
M3	F1	0.17	239	567	113	1198	4.0	10	0.38	2130
M4	F2	0.17	239	567	113	1198	4.0	10	0.38	2130
M5	F2	0.25	239	567	113	1198	4.0	16	0.38	2136
M6	F2	0.33	239	567	113	1198	4.0	21	0.38	2141

[a] The raw density of fresh concrete was not measured, but calculated with a constant pore volume.

Table 2: Properties of the used fibres.

fibre type	material	length l [mm]	diameter d [mm]	aspect ratio l/d [-]	density ρ_k [kg/dm³]
F1	PVA	6.18	0.11	55	1.18
F2	PVA	6.52	0.08	77	1.21

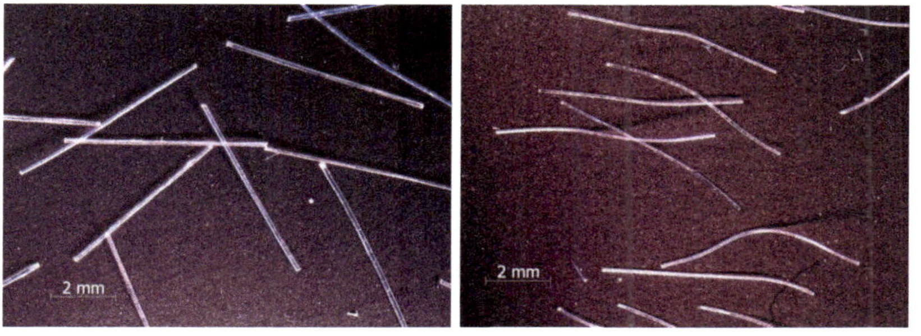

Figure 2: Optical differences PVA fibres (6.5x magnification). Left: F1. Right: F2.

2.2 Experimental set up

The aim of the investigations was to obtain information about the condition of fibres and maximum dosage targeting a successful injection. After adding water and achieving homogeneity, the sampling of the mixtures after certain times started. The experiments included the following tests:

- Rheological tests (at the times t = 15 min, 30 min, 45 min after water addition): Slump flow according to EN 1015-3 with Hägermann cone (SF) and measurements with Sliper device testing 4 load steps with 3 repetitions. With exception of M1 all mixes were remixed before testing.

- Mechanical properties (average of 3 to 6 values): dry density (ρ_d) – EN 12390-7, flexural strength (f_{ctm}) – EN 12390-5 and compressive strength (f_{cm}) –EN 12390-3/B1 each after t = 7 d, 14 d, 28 d. Test specimens were 3 prisms for each mix with dimensions a x b x c = 4 cm x 4 cm x 16 cm stored wrapped in foil under climatic conditions (RH = 65 %, T = 20 °C).

3 Experimental results and discussion

3.1 Rheological investigations

The individual results of the fresh concrete testing lists Table 3. The results of the Sliper experiments were not comparable to previous results3 in [14], so no direct correlation regarding injectability was possible4. Thus the focus of the evaluation lies within the test series. The relative slump flow according to Hägermann (Γ_P) stretched from $\Gamma_P = 0$ to 9.24 (M2 and M0). In general, with increasing time and increasing fibre volume the spread decreased.

Table 3: Results of fresh concrete properties in dependence of time.

mix	fibres dosage *f* [vol.-%]	Time *t* [min]	*SF* [mm]	Γ_P [-]	*a* [Pa]	*b* [Pa h/m³]
		15	320.0	9.24	126	255
M0	-	30	282.5	6.98	199	183
		45	255.0	5.50	88	244
		15	110.0	0.21	151	224
M1	0.33	30	100.0	0.00	175	276
		45	100.0	0.00	221	293
		15	100.0	0.00	348	331
M2	0.66	30	100.0	0.00	11	489
		45	100.0	0.00	723	122
		15	245.0	5.00	106	297
M3	0.17	30	260.0	5.76	53	342
		45	285.0	7.12	206	196
		15	230.0	4.29	86	216
M4	0.17	30	260.0	5.76	73	212
		45	252.5	5.38	246	125
		15	165.0	1.72	83	284
M5	0.25	30	165.0	1.72	277	177
		45	140.0	0.96	276	236
		15	122.5	0.50	198	210
M6	0.33	30	112.5	0.27	179	229
		45	130.0	0.69	232	237

SF slump flow (Hägermann cone) Γ_P relative slump flow (calculated by values *SF*)
a exp. value ≈ yield stress τ_0 *b* exp. value ≈ plastic viscosity μ

[3] Between M0 and SCM 1 the rheological parameters differed with $\Delta a = 150$ Pa and $\Delta b = 80$ Pa h/m³.

[4] Potential reasons are the different charges of the raw materials as well as the influence of the large filling of the dry mortar mix M0.

Starting at fibre contents of $f \geq 0.33$ vol.-% the relative slump flows decreased to $\Gamma_P \rightarrow 0$, with small improvements of fibre type F2. The rheological results of the Sliper device (parameters a and b) need discussion. The measurements require shearing of the lubricating layer. The increasing fibre content hindered the shearing leading to over proportionally high results. In general, the mixtures with a high spread had low yield stress ($a \approx \tau_0 = 11$ to 723 Pa, $a_m = 189$ Pa). This tendency was not so pronounced but also existent regarding the plastic viscosity ($b \approx \mu = 122$ to 489 Pa h/m³, $b_m = 240$ Pa h/m³). The values of the plastic viscosity scattered less (coefficient of variation $CV_a = 77$ %, $CV_b = 35$ %).

The needed pressure at different flow rates shows Figure 3. The bases for the curves were the Sliper tests. The individual tests (strokes) resulted in the shown points, a linear regression stated the dependence between pressure and different flows. With the exception of M2, the p-Q curves laid close together and differed with increasing time. The individual tests of M2 spanned the largest range at all measuring times and represented the maximum.

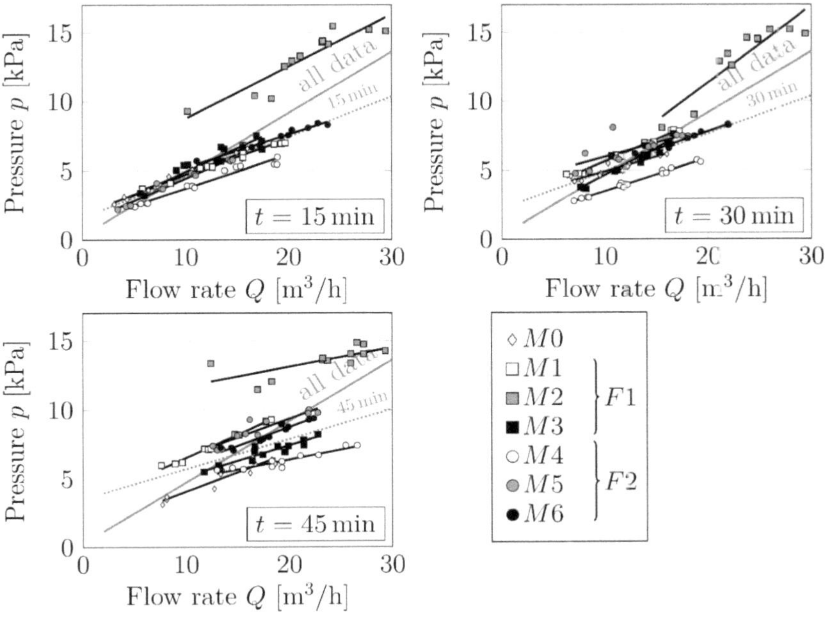

Figure 3: Calculation (Sliper) of pressure at flow rate at different times after water addition.

With growing time, the measuring ranges shifted towards higher pressures at higher flow rates. A rising fibre content led to an increase in the necessary pressure for the same flow rate. This was particularly noticeable for the ranges $t = 15$ min and partly $t = 30$ min. Starting at $t = 30$ min, the necessary pressures were lower for mixtures with more fibres (e.g. M6 compared to M5). The linear regressions (excepting M2) after $t = 15$ min and $t = 30$ min did not differ from each other and showed a smaller slope than the linear regression of all data. However, the slope after $t = 45$ min decreased significantly. The ascent of the curves in general is linked to the plastic viscosity. The decrease in plastic viscosity with increasing time could also be related to the interaction between fibre content and lubricating layer. Figure 4 illustrates the relationships between yield stress, plastic viscosity and relative slump flow, relative to the reference mix M0 after a corresponding measuring time. The diagrams differ according to fibre content, fibre type and time. The evaluation of the results regarding plastic viscosity and yield stress was most informative.

With the exception of M2, the mixtures were in the region of the reference at time $t = 15$ min. Ether the addition of fibres reduced the yield point and the plastic viscosity increased (M3, M5) or vice versa (M4, M6). The mixture M4 improved with regard to both parameters. At time $t = 30$ min the properties shifted towards increasing plastic viscosity. The range $t = 45$ min drifted in exactly the opposite way. The values shifted from the range $t = 15$ min in the direction of increasing yield stress. M2 repeated this pattern with more extreme values: The value after $t = 15$ min together with the origo formed a mirror axis, whereas after $t = 30$ min the plastic viscosity increased with a simultaneous decrease from the yield stress. After $t = 45$ min the values changed over proportionally. The values of parameter a_i/a_{M0} scattered the most in the comparison of both fibre types[5] and on average F2 was higher ($\Delta a_{F1,F2} = 0.15$). The relative results of b_i/b_{M0} tended to higher values at F1 with $\Delta b_{F1,F2} = 0.31$. The evaluation according to the relative slump flow $\Gamma_{P,i}/\Gamma_{P,M0}$ showed no differences ($\Delta\Gamma_{P,F1,F2} = 0.03$). With regard to the relative slump flow, the values dispersed increasingly over time.

[5] The mixtures M2 and M5 were not considered because the corresponding mixture with the same dosage was not available.

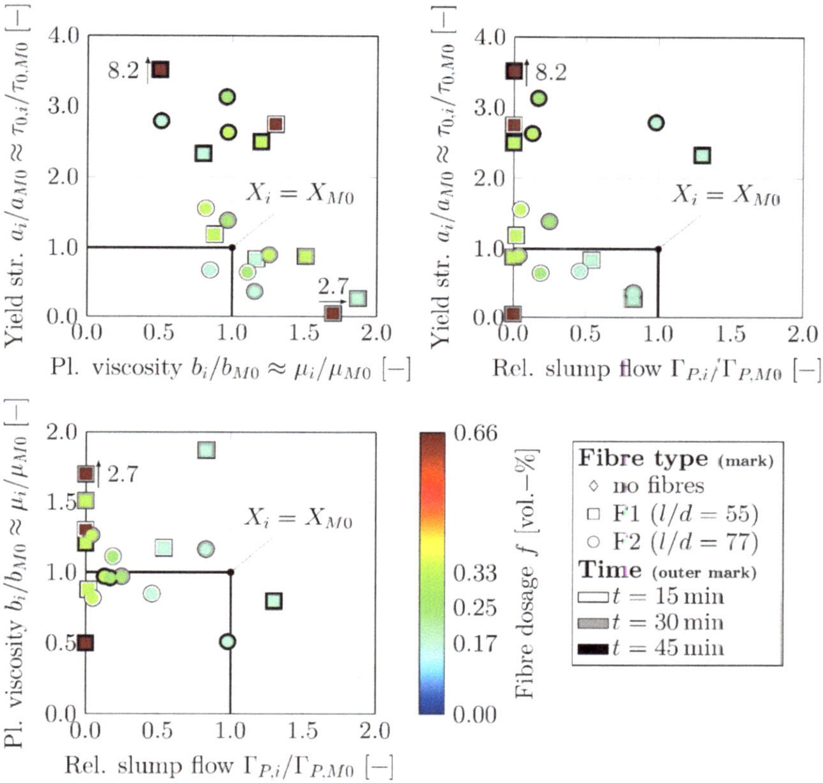

Figure 4: Measured rheological properties as a function of fibre type, fibre dosage and time.

3.2 Hardened concrete properties

Table 4 lists the hardened concrete properties of the mixes after the various test times. The mean average values of the dry density ρ_d as well as the flexural strength f_{ctk} were a summary of three individual values.

Six single results lead to the characteristic compressive strength f_{ck}. After a concrete age of $t = 7$ d f_{ctk} changed only slightly (exception M2), the compressive strength increased by an average of $\Delta f_{cm} \approx 20$ % up to $t = 28$ d. The mean value of the f_{ctk} for F2 decreased by $\Delta f_{ctk} = 0.86$ N/mm², but the coefficient of variation was also 3% lower with $CV_{fctk,F2} = 7$ %. With regard to f_{ck}, F2 also had lower mean values. The influence of fibre dosage in relation to the reference mix M0 illustrates Figure 5. All mixes except M2 showed a decrease in flexural strength after $t = 7$ d, only to return to the initial level after 28d. Only three compounds (M1, M2 and minimum M6) improved the reference mix after $t = 28$ d. The development of the compressive strength was contrary to this. All mixes demonstrated a clear improvement compared to the reference mix along with maxima after $t = 7$ d. After $t = 28$ d, the differences tended back towards the values of $t = 7$ d. With the exception of M2, improvements in the hardened concrete properties were slight for the mixes (M1, M3, M4, M6), and for M5 the compressive strength even decreased.

Table 4: Mean (density, n = 3) and characteristic hardened concrete properties (flexural with n = 3 and compressive strength with n = 6) of the mixes and their standard deviations.

mix	fibre dosage f [vol.-%]	age t [d]	density ρ_d [kg/dm³]	f_{ctk} [N/mm²]	σ_{fctk} [N/mm²]	f_{ck} [N/mm²]	σ_{fck} [N/mm²]
		7	2.3	8.3	0.4	69.0	1.1
M0	-	14	2.4	10.5	0.1	65.7	4.7
		28	2.3	5.5	2.1	78.9	4.2
		7	2.2	8.5	1.0	65.9	2.1
M1	0.33	14	2.2	9.4	0.4	70.7	4.2
		28	2.2	10.9	0.3	84.0	3.0
		7	2.2	10.2	1.8	76.0	2.2
M2	0.66	14	2.2	13.8	1.0	89.8	1.7
		28	2.3	8.5	1.9	98.7	2.3
		7	2.2	8.0	0.6	63.5	2.4
M3	0.17	14	2.2	7.1	0.9	77.9	3.0
		28	2.3	9.4	0.3	83.0	3.6
		7	2.3	9.0	0.1	70.2	1.9
M4	0.17	14	2.2	7.5	1.0	75.4	3.2
		28	-	8.6	0.4	83.0	3.6
		7	2.2	8.8	1.0	58.3	5.1
M5	0.25	14	2.2	7.8	0.9	69.7	3.8
		28	-	8.9	0.2	55.9	8.6
		7	2.2	10.2	0.0	63.0	3.6
M6	0.33	14	2.2	8.1	0.9	73.0	3.7
		28	-	9.0	0.5	82.6	2.8

f_{ctk} characteristic flexural strength σ_{fctk} standard deviation f_{ctk}
f_{ck} characteristic compressive strength σ_{fck} standard deviation f_{ck}

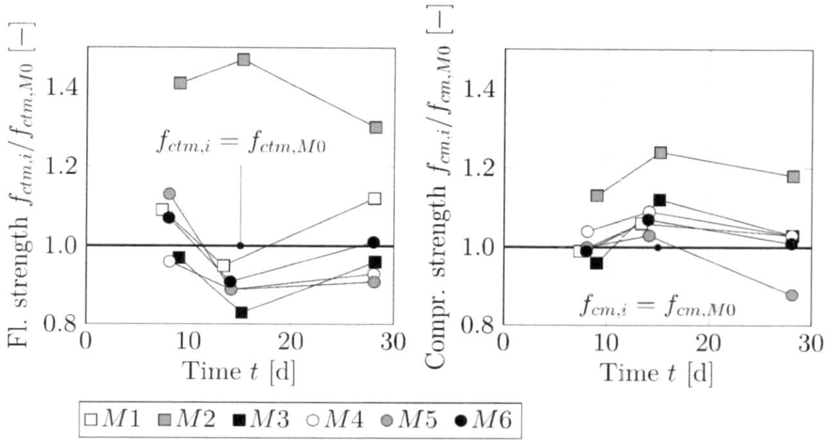

Figure 5: Temporal development of the flexural strength (left) and compressive strength (right) of the mixes in relation to the reference mix (M0).

4 Conclusion

The aim of the investigations was to obtain information about the condition of fibres and maximum dosage in relation to a successful injection. The tests included rheological investigations and hardened concrete tests. Seven mixes with fibre dosages of two PVA-fibre types (F1 and F2) from $f = 0.00$ vol.-% – 0.66 vol.-% were in focus. The rheological tests using the Sliper device and Hägermann cone were carried out after $t = 15$ min, 30 min and 45 min. Specimens tested after $t = 7$ d, 14 d and 28 d pointed out the development of the hardened concrete properties.

The characteristics of the fibres had a great influence on the fresh and hardened concrete characteristics. The increased number of fibres of F2 at the same fibre volume led to a homogenization of the rheological behaviour and a decrease of the hardened concrete performance. At the same time, the rheological tests of high-performance fine-grained fibre mortars revealed challenges. The fibres seemed to disturb the formation of the lubricating layer and lead to partly questionable results. Further investigations regarding the interactions between fibre dosages and lubricating layer would be interesting.

An advantage of the injection method is the short application time, a workability of $t = 45$ min was not necessary at all according to previous tests. When considering the time of $t = 15$ min, fibre contents between $f = 0.25$ vol.-% – 0.33 vol.-% seem to be the maximum for both fibre types with the starting materials used.

Acknowledgements

This work is a part of the project "PolymerConcreteLink plus – Sensor-activated pump adapter for modern construction" funded by the Federal Ministry for Economic Affairs and Energy (BMWi) in Germany. The funding programme is "Central Innovation Programme for Medium-Sized Enterprises (ZIM)".

References

[1] König, G.; Tue, N. V.; Zink, M.: Hochleistungsbeton. Bemessung, Herstellung und Anwendung. Berlin: Ernst&Sohn, 2001. (in German)

[2] Kwan, A.K.H.; Ling, S. K.: Filler technology for improving robustness and reducing cementitious paste volume of SCC, Construction and Building Materials 153, pp. 875–885, 2017.

[3] Peter, N. K.: Lexikon Bautechnik. 15.000 Begriffsbestimmungen, Erläuterungen und Abkürzungen. Heidelberg: Müller, 2005. (in German)

[4] Kim, J. S.; Kwon, S. H.; Jang, K. P. et al.: Concrete pumping prediction considering different measurement of the rheological properties, Construction and Building Materials 171, pp. 493–503, 2018.

[5] Mechtcherine, V.; Nerella, V. N.; Kasten, K.: Testing pumpability of concrete using Sliding Pipe Rheometer, Construction and Building Materials 53, pp. 312–323, 2014.

[6] Feys, D.; Khayat, K. H.; Khatib, R.: How do concrete rheology, tribology, flow rate and pipe radius influence pumping pressure? Cement and Concrete Composites 66, pp. 38–46, 2016.

[7] Choi, M. S.; Kim, Y. J.; Kim, J. K.: Prediction of Concrete Pumping Using Various Rheological Models, International Journal of Concrete Structures and Materials 8, Issue 4, pp. 269–278, 2014.

[8] Feys, D.; Khayat, K. H.; Perez-Schell, A. et al.: Prediction of pumping pressure by means of new tribometer for highly-workable concrete, Cement and Concrete Composites 57, pp. 102–115, 2015.

[9] Secrieru, E.; Fataei, S.; Schröfl, C. et al.: Study on concrete pumpability combining different laboratory tools and linkage to rheology, Construction and Building Materials 144, pp. 451–461, 2017.

[10] Ghanbari, A.; Karihaloo, B. L.: Prediction of the plastic viscosity of self-compacting steel fibre reinforced concrete, Cement and Concrete Research 39, Issue 12, pp. 1209–1216, 2009.

[11] Gerland, F.; Schleiting, M.; Schomberg, T. et al.: The Effect of Fiber Geometry and Concentration on the Flow Properties of UHPC. In: Mechtcherine, V.; Khayat, K.; Secrieru, E. (ed.): Rheology and processing of construction materials. RheoCon2 & SCC9. RILEM bookseries, volume 23, pp. 482–490. Cham: Springer, 2020.

[12] Li, M.; Li, V. C.: Rheology, fiber dispersion, and robust properties of Engineered Cementitious Composites, Materials and Structures 46, Issue 3, pp. 405–420, 2013.

[13] Hertwig, L.; Ulbricht, P.; Holschemacher, K.: Hochleistungs-Feinkornbetone im Injektionsverfahren. In: Greim, M.; Kusterle, W.; Teubert, O. (ed.): 27th Conference on the Rheology of Buildings Materials, Regensburg, pp. 174–185. Hamburg: tredition GmbH, 2018.

[14] Hertwig, L.; Holschemacher, K.: The injection process - effecting mechanical properties? In: Greim, M.; Kusterle, W.; Teubert, O. (ed.): 28th Conference on the Rheology of Buildings Materials, Regensburg, pp. 58–73. Hamburg: tredition GmbH, 2019.

Gründung des ZIM-Netzwerks

„Ressourcenschonende Recycling-Baustoffe der Zukunft“

	ZIM-Netzwerk
Vision	**Ressourcenschonende Recycling-Baustoffe der Zukunft**
Ziele der Netzwerke	Entwicklung von Baustoffen mit möglichst hohen Anteilen an Recyclingstoffen
Tasks	• Planung & Durchführung geförderter F&E-Projekte (Einzel- oder Kooperationsprojekte) • Schaffung eines positiven Bewusstseins für Abfallstoffe als neue Rohstoffe • Reduzierung Deponierungsbedarf für anfallenden Bauschutt • Steigerung der Nachhaltigkeit von Baustoffen der Gebäudehülle • Schonung bestehender Material- und Energieressourcen • Schaffung geschlossener Rohstoffkreisläufe
Mögliche Entwicklungsthemen	• Wiederverwendung von Rohstoffkomponenten aus bisher nicht trennbarem Bauschutt • Recycling von Gips, Beton und weiteren mineralischen Baustoffen • Rückführung sortenreiner Produktionsabfälle in den Produktionsprozess • uvm

Angestrebte Netzwerkpartner & Struktur:

mind. 6 Unternehmen (KMU)

assoziierte Industrieunternehmen

Forschungsinstitute, Hochschulen

Organisationen & Verbände

Netzwerkmanagement: RAS AG

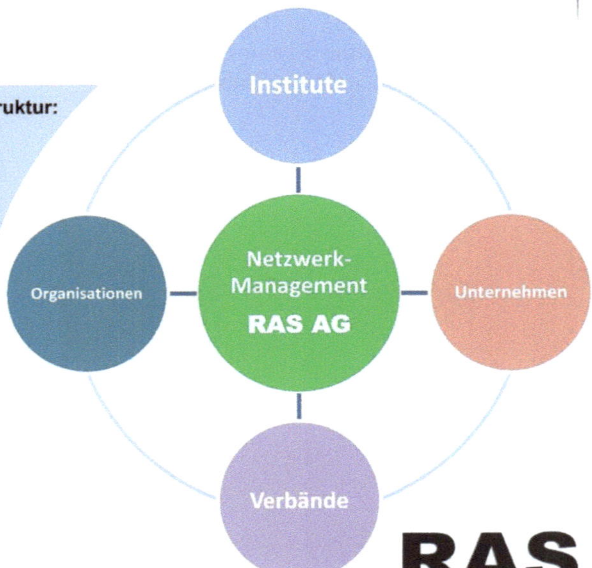

Ansprechpartner

RAS AG
An der Irler Höhe 3a
D-93055 Regensburg

Dr. Fabian Glaab
+49 (0)941 60 717 311
fg@ras-ag.com
www.ras-ag.com
www.rent-a-scientist.com

RAS
materials + technologies

Effect of Rheological Properties of Concrete Foundation on the Implementation of CFA-Piles

Yannick Vanhove, Chafika Djelal
Univ.Aartois, ULR 4515, Laboratoire de Génie Civil et géo-Environnement (LGCgE), Béthune, F-62400, France

Abstract

Pile-driving using a hollow stem auger with reinforcement cage (CFA-piles) has undergone widespread development over the past few years and is extensively practiced for building foundations. To the best of our knowledge, no current test or recommendation serves to guarantee an appropriate casting technique for lowering a pile without blocking the reinforcement cage from entering the fresh concrete. This type of concrete is cast without applying vibration, thus offering a highly fluid consistency. A concrete rest phase begins inside the boring cavity once concreting has been initiated; this phase lasts 30 minutes or longer, extending until the reinforcement cage has been completely lowered, during which time the concrete becomes restructured. The rest period will exert a direct impact on the ease of reinforcement cage placement. At the request of France's National Federation of Public Works (FNTP), a device has been developed to assess the capability of introducing reinforcements into concrete.

This study has been conducted to determine the reinforcement embedment capacity vs. pile casting time and the rheological behavior of concrete used to produce the specific pile. Tests were realized both in the laboratory and on site. The restructuring of concrete due to its thixotropy and loss of workability, alter the capacity of reinforcement to embed into the material, as evidenced by a decrease in such capacity throughout the study (for the given duration and concrete mixes under study).

It thus becomes possible to predict ultimate impediments when the extent of concrete restructuring and the sequence for lowering reinforcement cages at the jobsite are both known.

1. Introduction

When shallow foundations are infeasible due to the presence of compressible soils, deep foundations serve to transfer the structural loads of engineering facilities into higher quality layers found at greater depths. These processes are integrated into applications covering all building trades (housing construction, industrial, commercial, engineering structures, wind turbines, etc.). Pile-driving using a hollow stem auger, with or without reinforcements, has undergone widespread development over the past few years and is extensively practiced for building foundations [1,2,3]. This technique offers a quick (200 m per 10-hour shift) and economical solution, spared of noise and vibration nuisances. It is adapted to boring below the water table and moreover avoids the use of tubing, regardless of the type of ground traversed. Piles driven with a hollow stem auger are considered as both bored piles (extraction of a very small soil volume) and screw piles (lateral soil displacement due to pressurized concreting) [4]. This technique is widely practiced and found in 70% of all project sites requiring deep foundations.

The hollow augers are cast in situ through use of a continuous hollow stem auger. The auger (or worm drive) is bored into the soil at the calculated depth (Picture 1). It is then extracted in order to remove the bored soil, while the concrete is pumped and injected at low pressure via the auger shaft. The steel may be lowered into concrete by gravity or pushed gently to final position by hand. The reinforcing steel shall not be vibrated, driven, or otherwise guided into position by mechanical means.

Recommendations in conformity with the European standard EN 206-1, Fines content must be superior to 400 kg/m^3 end the concrete consistency must be either fluid (i.e. slump less than approx. 18 cm, spreading of 560 mm at 20°C for a pumped concrete) or plastic (slump above 15 cm, spreading of 500 mm for a poured concrete). The material must easily flow in between the reinforcements. Given the concrete casting period required to install piles, the consistency of the concrete must be preserved throughout the entire concreting phase.

Despite of these recommendations, in 15% of all cases, the reinforcement cage is difficult to install either at the beginning of its insertion into the concrete or over the last several meters. These difficulties are closely related to the rheological behavior of the concrete over time.

Picture 1: CFA pile-driving operations.

Fresh concrete undergoes various in situ phases during pile casting (Figure 1): transport of the concrete from the mixing plant to the site by means of a truck mixer; then concreting of the pile shaft by either pumping or gravity flow.

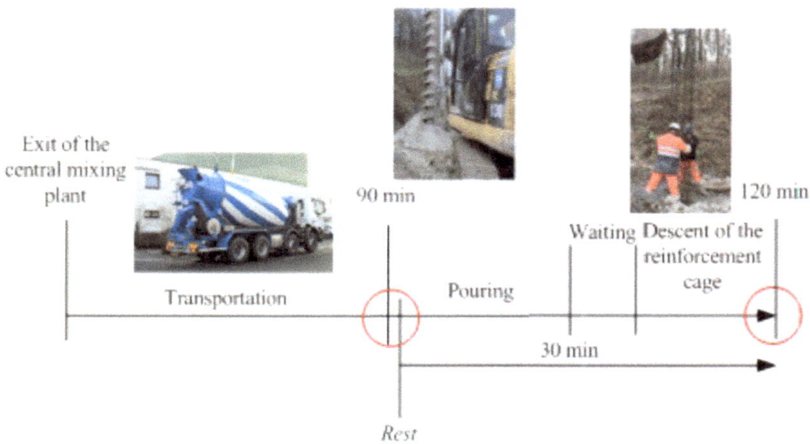

Figure 1: Concrete's life cycle on site.

During the concreting of the pile, the concrete is subjected to a high shear rate (on the order of 10 to 40 s^{-1}), which maintains the material in a disaggregated state. A concrete rest phase begins inside the boring cavity once concreting has been initiated; this phase lasts 30 minutes or longer, extending until the reinforcement cage has been completely lowered, during which time the concrete becomes restructured and the yield stress increase. The rest period will exert a direct impact on the ease of reinforcement cage placement.

More specifically, studies carried out on yield stress fluids, like concrete (cement pastes, Carbopol solutions, emulsions), reveal a correlation between object displacements in these fluids and their actual yield stresses [5, 6]. The goal of this study is thus to evaluate the influence of concrete rheology during the setting of reinforcement cages. A simple test of the capacity of reinforcements introduced into the concrete has been developed. A direct correlation may be drawn between the rheological properties and the penetration of the reinforcement cage over time of several common concrete mix design used for piles.

Restrictions are proposed taking into consideration concrete rheological properties over time regarding the in-situ phases during pile casting.

2. Experimental investigation

Before undertaking laboratory tests, a study was conducted on two piles construction sites in the north of France where the reinforcement cages have encountered some penetration problems into the concrete. Laboratory tests complete this study on seven common concrete mix designs used for piles.

2.1 Field investigations

The first site assessed is located in Lille (France) and dedicated to producing piles of 9.3 m high and 0.8 m in diameter. These piles were driven using a hollow stem auger by the firm Franki Fondations and the concrete was poured with a bucket. The concrete specified for this pile is presented in Table 1.

The second project site, located in Liévin (France), entails piles bored with a hollow stem auger by the firm Botte Fondations. The piles produced measure 1.2 m in diameter and 11 m in length. The concrete was pumped inside the allow stem auger. The concrete specified for this pile is presented in Table 2.

Table 1: Characteristics of investigated concrete in Lille

Designation	BPS NF EN 206/CN
Cement	CEMIII/A 42.5 N CE CP1 NF
Class of compressive strength	C30/37
Coarse aggregate	Maximal size of 20 mm
Slump value	200 mm
Environmental class	XF1-XC4-XD1
Adjuvant	MasterPolyheed 650 – 0.5% by cement mass

Table 2: Characteristics of investigated concrete in Lievin

Designation	BPS NF EN 206/CN
Cement	CEMIII/A 42.5 N CE CP1 NF
Class of compressive strength	C30/37
Coarse aggregate	Maximal size of 20 mm
Slump value	200 mm
Environmental class	XF1
Adjuvant	Plastiment 26 – 0.45% by cement mass

2.2 Test procedures with the reinforcement cage and the ICAR rheometer

The piles bored using hollow stem augers have a diameter ranging from 42 to 120 cm; their lengths may extend into the tens of meters. This experimental campaign has sought to reproduce in the laboratory a pile measuring 83 cm high and 17 cm in diameter, thus satisfying the construction principles set forth in the DTU 13.2 document (application of French standard) on piles. This dimensioning approximates the minimum pile diameter (42 cm). These piles have been reinforced by 5 HA14, with RL6 frames spaced every 20 cm (Picture 2).

The cavity in the soil is represented by a cylindrical tank with a 45-cm inner diameter. Edge effects have been considered. The cage coating measures 14 cm, i.e. 5 times greater than the maximum aggregate diameter (D_{max}=20 mm).

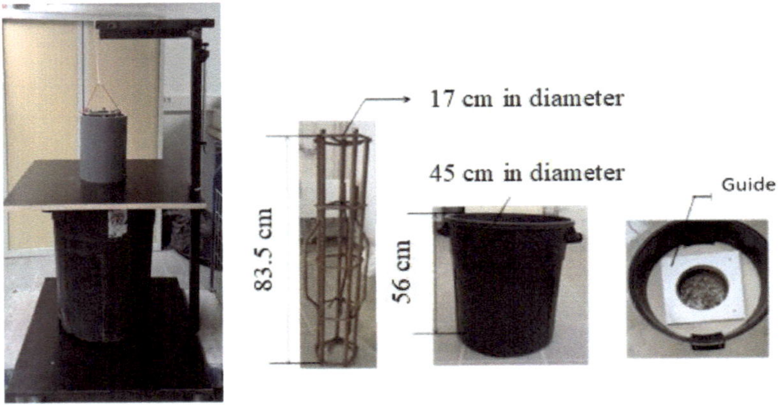

Picture 2: Test set-up to study the penetration of a reinforcement cage into concrete.

Concrete is introduced into the tank from a 30-cm height. The reinforcement cage is suspended at the level of the upper concrete surface via a metal cable, itself affixed to a jib crane braced by means of a solenoid. A PVC tube serves to guide the reinforcement cage during its movement. A switch releases the cage into the concrete, and its penetration can then be measured by implementing a displacement sensor. An acquisition system is programmed to record the reinforcement cage penetration height.

The protocol consists to introduce the concrete inside the bucket and to shear by mechanical energy the concrete to remove the concrete stress history. The reinforcement cage is also introduced into the concrete. After this initial test, the concrete is remixed and keep at rest during 30 minutes in the bucket. The test restarts 30 minutes later. The distance of penetration with time is recorded.

An ICAR rheometer, developed at the University of Texas, displays the advantages of being portable, simple and easy to use, and affordable [7]. This device comprises a lattice-like rotating tool for immersion into the concrete at a preset controlled rotational speed. Its operating principle is analogous to that of a coaxial cylinder rheometer.

The dynamic yield stress and the plastic viscosity are determined by applying a decreasing shear rate ramp (from 0.5 to 0.05 rps) to the concrete, with the rheometer then outputting the flow curve (change in concrete yield stress vs. imposed shear rate).

A thixotropy index Athix is evaluated according to Roussel's protocol [8], which offers the advantage of studying concrete behavior under shear for various rest times. The concrete pre-shear phase at 0.5 rps (much higher than the test value) removes the concrete stress history (the stress related to concrete placement in the rheometer tank), thus raising the level of test repeatability. This pre-shear is just realized after introducing the concrete inside the rheometer bucket. Thus, a constant shear rate threshold at 0.025 rps is applied for a long enough time to reach equilibrium. This shear rate is applied just after the pre-shearing and after a concrete resting time of 30 minutes. The impeller was kept inside the concrete during this rest time. The Athix is calculated at 30 minutes from the difference between the static yield stress (Pic value of the shear stress vs. time curve) at 0 and 30 minutes divided by the concrete resting time.

2.3 Materials and mixture proportioning

The cement used is CEM III A 42.5 N LH PMES CP1 NF, which proves to be particularly well adapted for foundation-related works; it is composed of 62% blast furnace slag and 35% clinker, containing 60% C_3S, 8.5% C_3A and 11% C_4AF (Table 3). Its Blaine specific surface area equals 4,200 m²/kg.

Table 3: Mixture proportion and characteristics of investigated concrete

Mixture	CF1	CF2	CF3	CF4	CF5	CF6	CF7
Cement, kg/m³	385	350	350	350	400	400	400
Water, kg/m³	166	190	190	190	166	166	166
Sand A 0/4 Cemex, kg/m³	853						
Sand B 0/4, kg/m³		410		410			
Sand C 0/4 Gand, kg/m³		415	415	415	829	829	829
Sand D 0/4 Gaurain, kg/m³			415				
Coarse aggregate A 4/20, kg/m³	885						
Coarse aggregate B 6/12, kg/m³		300	300	300	300	300	300
Coarse aggregate C 6/20, kg/m³		590	590	590	590	590	590
Admixture 1, % cement mass	1.2						
Admixture 2, % cement mass		0.8					
Admixture 3, % cement mass			0.8	0.8			
Admixture 4, % cement mass					0.35	1.15	0.89
VMA, % water					0.39	0.81	0.81
Slump, cm	22±2	23±1	21±1	23±1	24±2	23±1	21±1

The admixtures introduced were super-plasticizer high-range water reducer. Adding these admixtures then not only improves the workability of the concrete but extends its period of workability as well. The CF5, CF6 and CF7 mixtures were proportioned with a viscosity modifying admixture (VMA) to promote an increasing of the plastic viscosity and thixotropy of concrete mixtures CF2 to CF4. All mixtures were prepared in batches of 90 L using a pan mixer of 120 L capacity. The mixing sequence consisted of homogenizing the sand, coarse aggregate, and cement for 1 minute before introducing a part of the total water for 2 minutes more. The concrete was then mixed for 3 minutes with the rest of the water with admixture.

3. Experimental results and discussion

3.1 Rheological properties

Table 4 summarizes the rheological properties of the investigated mixtures. The dynamic yield stress, the static yield stress and the plastic viscosity were determined just after the mixing sequence. The static yield stress has been measured a second time after a concrete rest period of 30 minutes.

Table 4: Concrete rheological properties

Mixture	CF1	CF2	CF3	CF4	CF5	CF6	CF7	Lille	Lievin
Dynamic yield stress, Pa	23.3	213.5	406	173.6	216	227	197	217	202
Plastic viscosity, Pa.s	86.9	6	6.1	6.4	34.8	25.7	44.8	25.9	27.8
Static yield stress at 0 minutes, Pa	61	246	302	142	742	732	570	442	455
Static yield stress at 30 minutes, Pa	133	813	687	429	2612	2803	2253	1179	1032
Athix, Pa/s	0.04	0.31	0.21	0.16	1.04	1.15	0.94	0.41	0.32

3.2 Insertion of the reinforcement cage into wet concrete

Typical variation of the cage penetration inside the concrete at 0 minutes and 30 minutes later is shown in Figure 2.

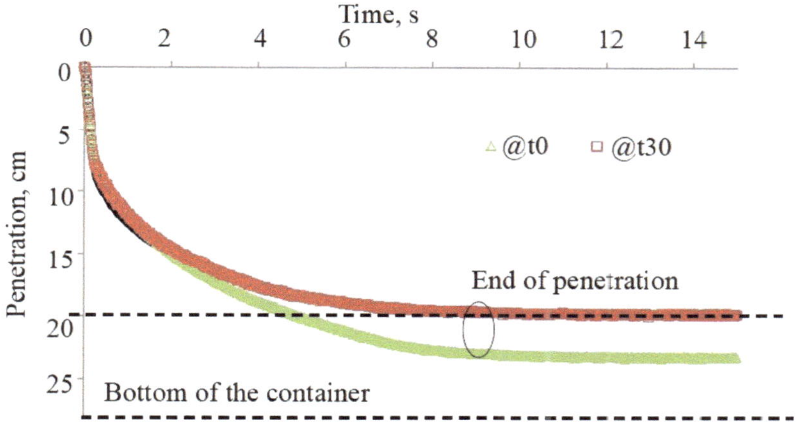

Figure 2: Evolution of the cage penetration with time at 0 and 30 minutes.

The cage penetration increases sharply during the first second of testing before slowing down due to the concrete rheological properties. Maximum penetration in this example occurred after approximately 9 seconds to reach a maximum penetration value of 23 cm for the test realized just after concrete mixing and 20 cm after a rest time of 30 minutes.

3.3 Effects of concrete rheological properties to reinforcement cage insertion capacity

End of cage penetration values determined after a concrete resting time of 30 minutes are compared in Figure 3 to static yield stress of the concrete at the same time.

As illustrated in Figure 3, the end of cage penetration values is shown to correspond well to the static yield stress values determined over the period of concrete at rest for 30 minutes. A stress threshold can therefore be found at approximately 800 Pa, from which point the cage penetration is constant. The existence of such a point presumes that beyond this stress level, the reinforcement cage jamming phenomenon is readily observed, in which case vibration would be required. The concrete develops reversible microstructural change at rest with time that results in shear stress increasing. These phenomena can be quantified from a thixotropy index named Athix [8]. The Athix values determined from the evolution of static yield stress are reported on the Figure 3. Regarding the static yield stress threshold value estimated at 800 Pa, a limit of Athix can be associated at 0.3 Pa/s. The respect of this limit value can contribute to avoid the reinforcement cage jamming phenomenon and particularly to give enough time to workers to introduce the reinforcement cage into the concrete.

The evolution of the static yield stress with time for Lievin's site and Lille's site is simulated with CFA pile-driving operations and reported in Figure 4 and 5, respectively.

At each CFA pile-driving operations (Figure 4), the yield stress increases to reach 800 Pa at around 18 minutes. The Athix value is equal to 0.32 Pa/s. A driving refusal has been observed at 10.2 meters when the static yield stress was equal to 725 Pa. Workers do not have enough time to completely lowered the reinforcement cage.

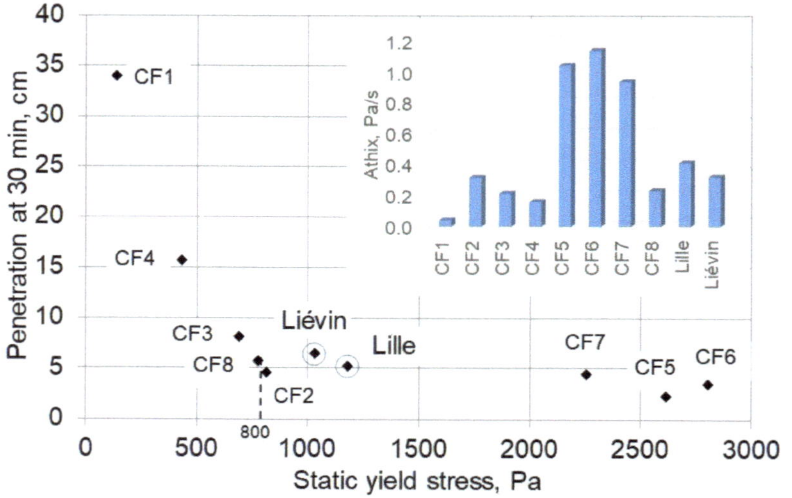

Figure 3: Variation of penetration at 30 minutes with the concrete static yield stress.

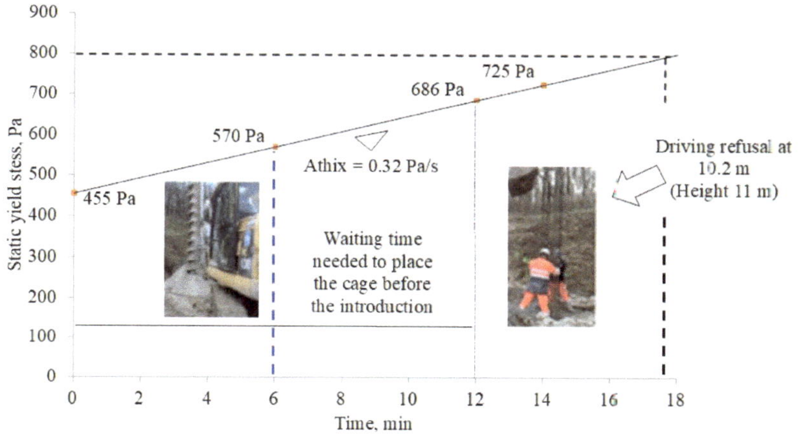

Figure 4: Evolution of the concrete static yield stress with time on Lievin's site.

Concerning Lille's site, the driving refusal appeared at 6.3 meters when the concrete static yield stress reached 800 Pa (Figure 5).

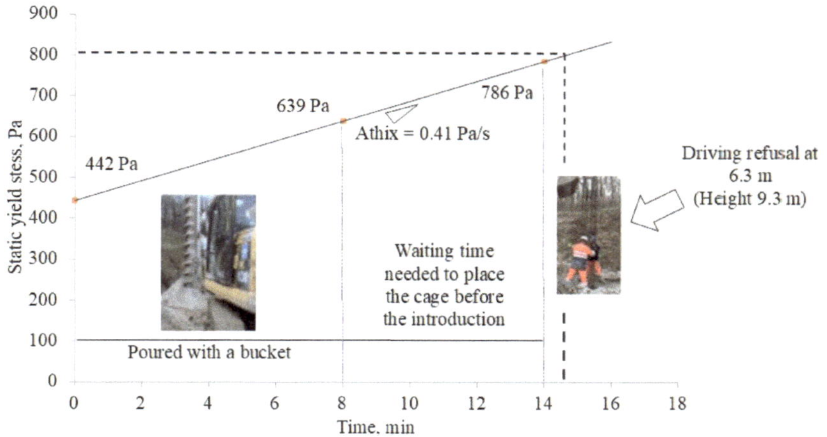

Figure 5: Evolution of the concrete static yield stress with time on Lille's site.

These two applications on site show that a lower thixotropy index (inferior to 0.3 Pa/s) would allow to workers a sufficient time to insert the reinforcement cage at the desired depth. This recommendation can avoid the use of vibratory drive at the head which is commonly practice in Europe to install the cage. The use of a vibratory drive head could lead to problems with cages that are not securely tied or welded and could also produce segregation and bleeding if the concrete mix is not well proportioned.

4. Conclusions

The study presented herein evaluates the effect of time at rest on both concrete rheology and the capacity of reinforcements to penetrate the concrete during CFA pile-driving operations. It has also demonstrated that the alteration kinetics of concrete at rest exert an influence on the capacities of reinforcements to become embedded. Based on the results of the study, the following conclusion can be drawn:

- The rheological properties obtained using the ICAR rheometer have served to corroborate observations recorded with the cage penetration setup. This tool offers extremely high repeatability. It is also sensitive to sampling variations in concrete at the mixing truck discharge.
- A very strong correlation can be observed between the time when reinforcement cage progression in the concrete ceases and the static yield stress.
- A thixotropy index lower than 0.3 Pa/s would allow workers to have enough time to insert the reinforcement cage at the desired depth.
- CFA pile-driving operations could be improved by using a guide to decrease the time needed to place the cage before its introduction and to take advantage of the gravity force of the cage.

5. Acknowledgements

This effort could not have been accomplished without the valuable assistance of all project participants. We would like to express our sincere gratitude to the FNTP Federation, Holcim Cement (Lumbres), Holcim Aggregates (North), for their contributions.

Bibliography

[1] Norman Mure; Jason N. Scott; Derek Seward, Stephen D Quayle; Chris R.I. Clayton; M. Rust: Just-in-Time Continuous Flight Auger Piles Using an Instrumented Auger, Proceedings of the 19th ISARC, I.A.A.R.C., 455-460, Washington, U.S.A.: IAARC, 2020

[2] Tarek Zayed: Productivity and Cost Assessment for Continuous Flight Auger Piles, Journal of Construction Engineering and Management, 677-689, ASCE, 2005

[3] Fathi M. Abdrabbo; Khaled E. Gaaver: Installation effects of auger cast-in-place piles, Alexandria Engineering Journal, 281-292, Elsevier, 2012

[4] Martin D. Larisch: Current Practice of CFA Piling an Austria and New Zeland, International Conference on Deep Foundations and Grout Improvement, 572-580, DFI/EFFC, Rome, Italy, 2018

[5] Didier Loutens; Pierre Jousset; laetitia Martinie; Nicolas Roussel; Robert J. Flatt: Yield stress during setting of cement pastes from penetration test, Cement and Concrete Research, 401-408, Elsevier, 2009

[6] Peter H.T. Uhlherr; Gian Guo; Tunan Fang; Carlos Tiu: Static measurement of yield stress using a cylindrical penetrometer, Korea Australia Rheology Journal, 17-23, Springer, 2002

[7] Eric P. Koehler; David W. Fowler: Development of a portable rheometer for fresh Portland cement concrete, Research report International center of Aggregates Research, 105-3F, 2004

[8] Nicolas Roussel: A thixotropy model for fresh fluid concretes: theory, validation and applications, Cement and Concrete Research, 1797-1806, Elsevier, 2006.

MASTERS OF BLENDING KNIELE

Der Mischer für klassische Frischbetonversuche

- Ausbreitmaß
- Setzfließmaß
- VdZ-Trichter
- LCPC-Box

Mit dem integrierten **Rheometer** und **Tribometer** ermöglicht der KKM-RT die Eigenschaften des Mischguts in relativen und absoluten Einheiten zu messen.

Labor – Intensivmischer

KKM – RT 15 / 22,5

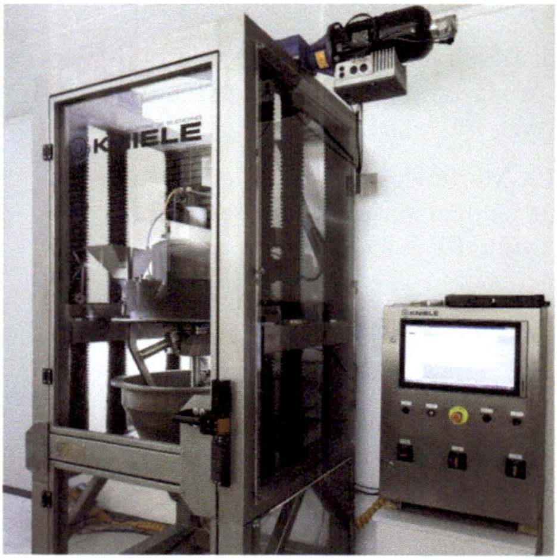

Features:

Rheometer und Tribometer

- Messung absolut und relativ

Spezialrührwerk

- Aufschluss von Desagglomeraten

Mischersonde

- Messung der Feuchte

Kamera

- Beobachtung des Mischvorgangs

Rheometer mit Schnellwechselsystem

Spezialrührwerk

Mischersonde

Laborversuche mit Schaumbetonen

Kniele GmbH
Gemeindebeunden 6
88422 Bad Buchau
FON: +49 7582 9303 - 0
FAX: +49 7582 9303 - 30
Mail: info@kniele.de

blending • dosing • conveying – made in Germany

kniele.de

Experimental setup to determine rheological properties of frazil ice

Felix Paul, Tommy Mielke, Doru C. Lupascu
Institute for Materials Science and Center for Nanointegration Duisburg-Essen (CENIDE), University of Duisburg-Essen, Germany, felix.paul@uni-due.de, www.uni-due.de/materials/

Abstract

Frazil ice is the very first ice that forms in the marginal ice zone of the Antarctic. The rheological properties of frazil ice play a significant role in numerical calculations of sea ice behavior. Up to now no in situ measurements of the rheological properties of frazil ice in the Marginal Ice Zone of Antarctic have been conducted. The trajectory of the 2019 SCALE Winter Cruise and a test setup for measuring rheological properties in the Marginal Ice Zone of the Antarctic are presented. A description of a sampler to get undisturbed samples of frazil ice for rheological measurements is explained. The eBT-V rheometer and the modified vane are presented. Several rheological measurements were conducted in the Antarctic during the Winter Cruise 2019 on the SA Agulhas II.

1 Introduction

One of the largest seasonal changes on earth is the growth and decay of sea ice in the Antarctic polar region [1]. These seasonal variations are of great importance for the flora and fauna of Antarctica and for the global climate. Frazil ice consists of loose disc-shaped ice crystals. It is formed in turbulent and supercooled water [2] and plays an important role in the freezing process in the marginal ice zone of the Antarctic. The marginal ice zone is the transition zone between consolidated ice and open water, where the sea ice concentration lies in between 15 % and 80 % [3]. When a sufficiently large quantity of frazil ice is present, it clusters to form cakes of ice, which later develop into pancake ice. These pancakes grow and a frazil/ pancake mixture is created (Picture 1). Subsequently, pancakes are rafted and freeze together. If the waves are sufficiently dampened by this frazil/pancake layer, a closed ice cover can form.

Picture 1: Frazil/ pancake ice mixture in front of the SA Agulhas II.

This study investigates a test setup to determine the rheological properties of the frazil ice described above. Rheological properties of frazil ice are important for numerical simulations of sea ice in the marginal ice zone of the Antarctic. These values are necessary to model the interaction between frazil ice, pancake ice and water. Frazil ice acts like a viscous fluid, which works as a damping mechanism between the pancakes. No directly measured rheological properties of frazil ice are available in literature. Only rheological measurements in wave tanks have been recorded [4]–[6]. Therefore, this study focused specifically on in-situ data collection during the Winter Cruise in 2019 in the Antarctic.

2 Frazil ice

Frazil ice crystals are simple in shape, appearing as thin, circular, disc-shaped crystals. They tend to have a diameter between 0.01 to 5 mm and the thickness varies between 1 to 100 μm [7]–[9]. Frazil ice is formed through a mass-exchange process, which takes place at the air-water interface. In 1974, Osterkamp [10] studied the surface of a supercooled freshwater river in order to discover a potential nucleation process, for frazil ice. He proved that ice crystals falling into the water could initiate nucleation. During the experiments carried out by Osterkamp, the water temperature of the river was between - 0.012 °C and - 0.050 °C, thus the conditions of the river were supercooled. Airborne ice crystals were observed during the investigation with flashlights. At temperatures below – 8 °C ice crystals were discovered in the air above the river, whereas this did not occur during warmer temperatures. Once the river froze, there were no more ice crystals within the air above the river. Therefore the crystals of the primary nucleation must originate from the water and serve as a nuclei for frazil ice formation, as soon as they fall back into the water. [10] When enough frazil crystals are formed, a secondary nucleation process takes place. The secondary nucleation process is a homogeneous nucleation where ice crystals from collision are the initial crystallization nuclei [9], [11].

Therefore supercooling and turbulence of the water are necessary for frazil ice production. A liquid is supercooled below its freezing point without freezing. A typical time temperature profile during the formation of frazil ice in fresh water is presented in Figure 1.

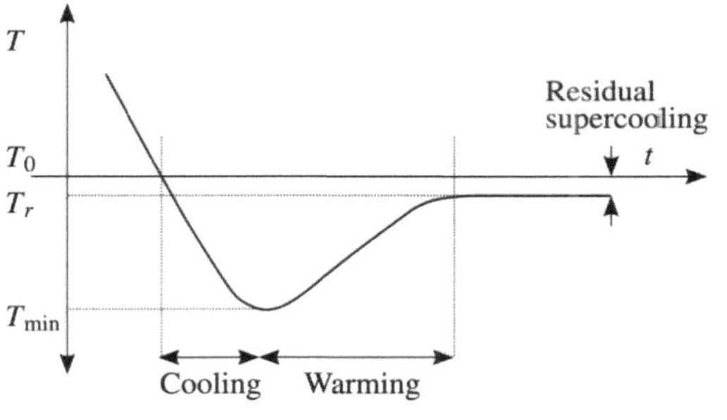

Figure 1: Typical supercooling of fresh water after Ye [12].

Until the minimum temperature T_{min} is reached, the water has a net heat loss due to the cold air on the surface of the water. Shortly before the point at which minimum temperature is reached, the liquid is seeded with nuclei and ice platelets begin to form. Newly formed ice crystals release a specific amount of heat in the water and increases its temperature. The more ice platelets start to form, the more heat is released into the water until the equilibrium between heat loss and heat gain is reached [9]. Equilibrium is reached at the point of minimum temperature. Thereafter, the period of warming begins, since the energy gained from the crystal formation is higher than the heat loss through the cold air. The region of residual supercooling is also known as the principal period of supercooling and is an important force for the ongoing growth of frazil. If the residual temperature is equal to zero, there will not be any frazil production in the water [12].

Turbulence is the second important prerequisite for frazil ice production. The minimum turbulence for the production of frazil ice varies in literature. The minimum flow velocity in the experiment by Hanley [13] is 15 cm/s, 9.1 cm/s by Smedsrud [14], 24 cm/s by Hanley [15] and 33 cm/s by Carstens [16] to name just some examples. A stronger intensity in turbulence increases the rate of frazil ice growth, which corresponds to a rise in the rate of change of water temperature. This increase in frazil ice growth can be explained by an improved heat transfer between the water and frazil ice, as well as by a higher collision rate between the frazil ice platelets [17].

3 Experimental set-up

The aim of the experimental set-up is to get an undisturbed frazil ice sample from the top layer of the ocean. The experiments were conducted on the South African research ship SA Agulhas II. The SCALE cruises are funded by the South African National Research Foundation (NRF) through the South African National Antarctic Programme (SANAP), with contributions from the Department of Science and Innovation and the Department of Environmental Affairs. The frazil ice was lifted out of the water with a self-constructed frazil sampler and the viscosity of the specimen was measured with the eBT-V Rheometer from Schleibinger.

The frazil sampler (Picture 2) is designed to retrieve an undisturbed frazil ice and water specimen from seawater. It consists of a metal cylinder (Picture 2 (a)), which can move on four steel bars (Picture 2 (b)) between two metal plates (Picture 2 (c)). The metal cylinder acts as the wall of a bucket. A foam mat (Picture 2 (d)) attached to the upper side of the bottom metal plate acts as a seal when the cylinder is placed on top, forming a water-tight "bucket", preventing water from draining out. The frazil sampler can be opened and closed by pulling either of the two ropes. Pulling the green (Picture 2 (e)) rope moves the cylinder to the top and opens the frazil sampler.

By releasing the green rope and pulling on the red (Picture 2 (f)) rope the bucket glides to the bottom plate and seals the sampler. The force from the red rope presses the metal cylinder against the foam mat and closes it, making it water-tight. The red rope is thread through a hole to the underside of the metal and looped back through yet another. Between the loop on the underside, a 'semi-cylindrical metal roll' (Picture 2 (g)) is placed to reduce the friction between the rope and the metal plate. The ends of each rope are thread through a metal bolt (Picture 2 (h)) attached to the cylinder. To fasten the ends of the rope, 2 screws are screwed into the metal bolt, holding the rope in place, this prevents movement and reduces frictional damage. The entire weight of the frazil sampler hangs on the metal hook (Picture 2 (i)) in the center of the top plate and is not carried by the green and red rope. The yellow mark (Picture 2 (j)) on the steel bars shows how deep the frazil sampler has to be immersed into the water. There are four smaller hooks (Picture 2 (k)) on the edges of the top plate. They are used to attach ropes to them and hold the frazil collector in position during overboard operation. It prevents the frazil sampler from rotating and swinging around.

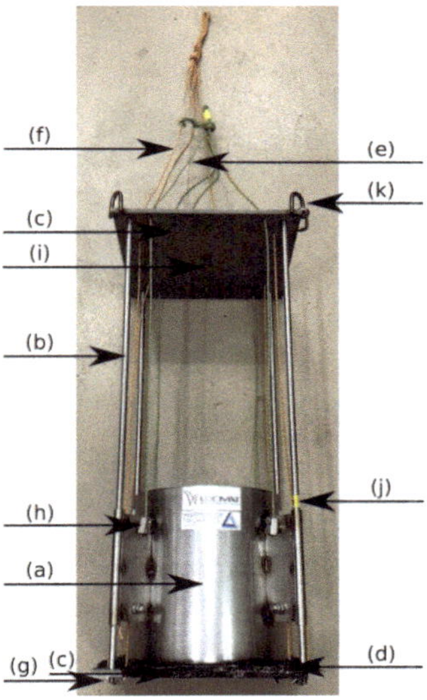

Picture 2: Frazil sampler to receive undisturbed samples of frazil ice.

The SA Agulhas II is a South African vessel, ice class IACS PC 5, funded by the Department of Environmental Affairs of South Africa. The keel of the vessel was laid in 2011. It has a gross tonnage of 12897.00 t, overall length of 134.00 m, a range of up to 27000 km and is powered by four main engines. The vessel provides many research possibilities by offering a wide variety of equipment. Several laboratories, cranes, under water tests, a great crew and a lot of more configurations allow for a good research environment. One of these implements is the A-frame. The A-frame is located starboard and allows to lift test equipment outside the ship. The height of the A-frame is 4.79 m and the arm is 2.00 m long.

To use the A-frame, the door (Picture 3 (e)) is opened automatically. Then the whole A-frame can be swung out. The experimental setup can either be attached to the small roll (Picture 3 (c)) or to the big black roll next to it. Both of them are connected to winches. The green and red rope from Picture 2 can be looped over the carabiner on the right side of the A-frame (Picture 3 (f)). To hold the frazil collector in position two operators can stand behind the barrier on the left and right sides (Picture 3 (a)).

Picture 3: A-frame of the SA Agulhas II.

The eBT-V by Schleibinger (Picture 4) is a rheometer originally designed for fresh concrete. It is a portable, compact rheometer and can be controlled via a smartphone with Bluetooth connection. As the rheometer is battery driven it well suits the requirements for on-board use. It can be operated in two different modes. The P-mode which is less suitable for the rheological analysis of frazil ice and the V-mode [18].

The vane delivered with the device has the dimensions R_V=32.5 mm and h=75 mm. To increase the torque measured by the rheometer a larger vane was manufactured. The new dimensions (Picture 5) are R_V=51.5 mm and h=103 mm.

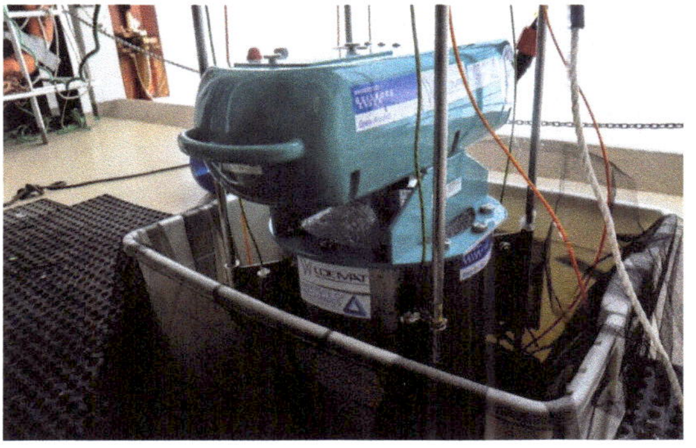

Picture 4: eBT-V Rheometer on the SA Agulhas II.

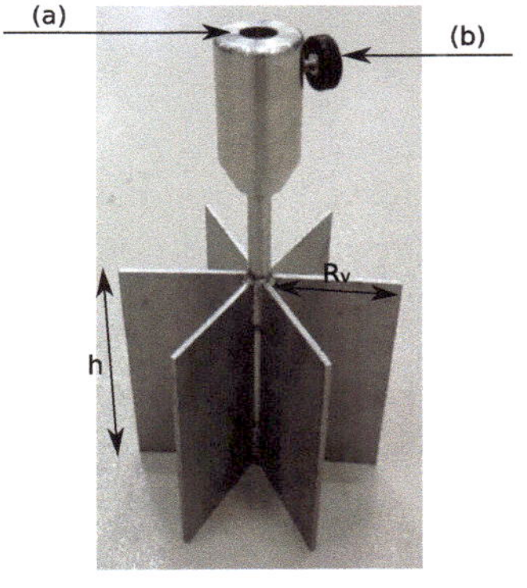

Picture 5: Vane for measuring frazil ice.

The black screw (Picture 5 (b)) is used to fasten and unfasten the vane on the device shaft.

The frazil ice analysis was conducted during the 2019 Winter Cruise on the SA Agulhas II. The cruise ran from 18.07.-08.08.2019. During the cruise several scientific groups conducted research in different fields such as biochemistry, trace elements and sea ice. In total, 94 scientists participated in the cruise. The trajectory of the cruise is shown in Picture 6. The cruise reached the ice edge, although this is not indicated on the map in Picture 6. The Antarctic continent has a landmass of about $14*10^6$ km^2 and is surrounded by the Southern Ocean. An area of about $3*10^6$ km^2 of the Southern Ocean is covered with sea ice in summer, whereas in winter, about $19*10^6$ km^2 is covered with sea ice [19].

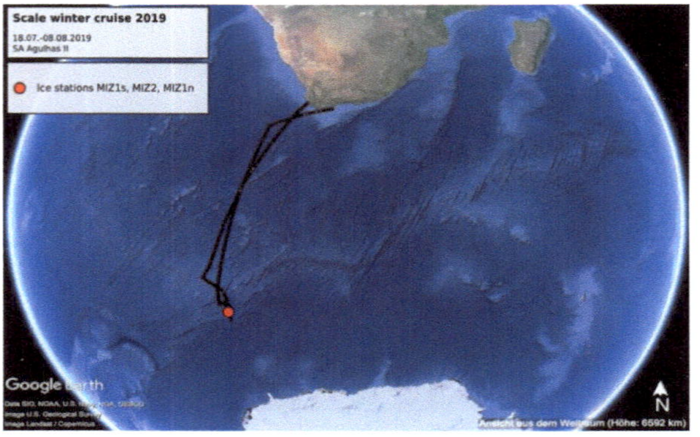

Picture 6: Trajectory of the SA Agulhas II during the Winter Cruise 2019.

3 Testing procedure

The aim of this project is to obtain an undisturbed frazil ice and water specimen from the ocean to measure the viscosity of the mixture. To collect frazil ice and water specimens, the hook of the frazil sampler (Picture 2 (i)) is connected to the winch of the A-frame (Picture 3 (c)). The frazil ice sampler is openend and then lowered to the water surface. The open frazil ice sampler is moved close to a gap between pancakes and is then lowered down to the yellow mark (Picture 2 (j)) touching the water. Frazil ice may be destroyed by turbulence as the frazil ice sampler is immersed into the water. To ensure that the sampler collects an undisturbed specimen, the sampler is moved slightly (either left or right). Thereafter the operator closes the frazil ice sampler and lifts it rapidly out of the water. On board of the ship a box is prepared to place the frazil ice sampler into it. A cotton sheet is spread over the box before placing the frazil sampler in the box on top of the cotton sheet. Frazil ice measurements commence once the sampler is placed in the box. Rheological measurements have to start quickly to prevent the frazil ice from melting or other structural changes. First, a small specimen of the water and frazil mixture is scooped out to determine salinity. Afterwards the temperature of the frazil ice and water mixture and the volume of the sample are measured. The eBT-V rheometer is placed into the frazil sampler with the water and frazil mixture and the measurement is started via a smartphone. After the viscosity measurement is recorded, the frazil and water mixture is separated into a pure frazil ice sample and a pure water sample. This is done by opening the frazil sampler and allowing the frazil and water mixture to flow through the cotton sheet into the box. The frazil sampler can now be used to collect another specimen.

The modified Schleibinger rheometer has proven to be a viable testing device for frazil ice studies. It will serve to further undisclose the secrets of our polar regions. Details of the rheological data and implication on ice formation will be presented elsewhere.

6 Acknowledgements

The SCALE cruises are funded by the South African National Research Foundation (NRF) through the South African National Antarctic Programme (SANAP), with contributions from the Department of Science and Innovation and the Department of Environmental Affairs.

We are grateful to Sebastian Skatulla, Jörg Schröder and Marcello Vichi for having opened up this wonderful opportunity for Antarctic ice research. Furthermore, we thank Riesna Audh, Mark Hambrock, Rutger Marquart, Carina Nisters, Tokoloho Rampai and the crew of the SA Agulhas II for assisting the experiments on the Winter Cruise 2019.

Bibliography

[1] M. J. Doble, "Pancake ice formation in the Weddell Sea," *J. Geophys. Res.*, vol. 108, no. C7, 2003.

[2] C. Petrich and H. Eicken, "Overview of sea ice growth and properties," in *Sea Ice*, 2017, pp. 1–41.

[3] M. Vichi *et al.*, "Effects of an Explosive Polar Cyclone Crossing the Antarctic Marginal Ice Zone," *Geophys. Res. Lett.*, pp. 1–24, 2019.

[4] K. Newyear, "Comparison of laboratory data with a viscous two-layer model of wave propagation in grease ice," vol. 104, pp. 7837–7840, 1999.

[5] R. Wang and H. H. Shen, "Experimental study on surface wave propagating through a grease-pancake ice mixture," *Cold Reg. Sci. Technol.*, vol. 61, no. 2–3, pp. 90–96, 2010.

[6] X. Zhao and H. H. Shen, "Wave propagation in frazil/pancake, pancake, and fragmented ice covers," *Cold Reg. Sci. Technol.*, vol. 113, pp. 71–80, 2015.

[7] S. Martin and P. Kauffman, "A Field and Laboratory Study of Wave Damping by Grease Ice," *J. Glaciol.*, vol. 27, no. 96, pp. 283–313, 1981.

[8] S. F. Daly, "International Association for Hydraulic Research Working Group on Thermal Regimes Report on Frazil Ice IAHR AIRH," no. August, 1994.

[9] V. McFarlane, M. Loewen, and F. Hicks, "Measurements of the evolution of frazil ice particle size distributions," *Cold Reg. Sci. Technol.*, vol. 120, pp. 45–55, 2015.

[10] T. E. Osterkamp, "Frazil-Ice nucleation by mass-exchange processes at the air-water interface," *J. Glaciol.*, vol. 19, no. 81, pp. 619–625, 1974.

[11] S. P. Clark and J. C. Doering, "Frazil flocculation and secondary nucleation in a counter-rotating flume," *Cold Reg. Sci. Technol.*, vol. 55, no. 2, pp. 221–229, 2009.

[12] S. Q. Ye, J. Doering, and H. T. Shen, "A laboratory study of frazil evolution in a counter-rotating flume," *Can. J. Civ. Eng.*, vol. 31, no. 6, pp. 899–914, 2004.

[13] T. O. D. Hanley and G. Tsang, "Formation and properties of frazil in saline water," *Cold Reg. Sci. Technol.*, vol. 8, no. 3, pp. 209–221, 1984.

[14] L. H. Smedsrud, "Frazil-ice entrainment of sediment: large-tank laboratory experiments," *J. Glaciol.*, vol. 47, no. 158, pp. 461–471, 2001.

[15] T. O. Hanley and B. Michel, "Laboratory formation of border ice and frazil slush," *Can. J. Civ. Eng.*, vol. 4, no. 2, pp. 153–160, 1977.

[16] T. Carstens, "Experiments with supercooling and ice formation in flowing water.," *Geofys. Publ.*, vol. 26, pp. 1–18, 1966.

[17] R. Ettema, M. F. Karim, and J. F. Kennedy, "Laboratory experiments on frazil ice growth in supercooled water," *Cold Reg. Sci. Technol.*, vol. 10, no. 1, pp. 43–58, 1984.

[18] S. Geräte, "Schleibinger eBT-V Rheometer for fresh concrete - Manual," 2017.

[19] S. Stammerjohn and T. Maksym, "Gaining (and losing) Antarctic sea ice: variability, trends and mechanisms," in *Sea Ice*, 2017, pp. 261–289.

Rheology - suitable measurement methods for building materials

Helena Keller, Markus Greim
Schleibinger Geräte Teubert und Greim GmbH, Buchbach, Germany, www.schleibinger.com

Introduction

The ever-growing demands on the properties of mortar and concrete are closely linked to the permanent further development of these cement-based mixtures. The research in these materials can be divide into three fields of investigation which are durability, dimension stability and workability. While durability and dimension stability focused on hardening processes, the workability deals with the liquid or plastic state of cementitious materials. The material properties such as hardness, strength and durability can be only achieved following a period of plasticity. Hence, the adjustment of the workability settings and the control of them play an essential role for the further properties of the materials.

The plastic state and thereby also the workability parameters such as pumpability, flowability, moulding, compaction and particularly self-compaction, and the ability of vent of concrete and mortars are related to the rheological properties of the fresh cement paste.

The properties of the fresh cement paste are strongly influenced by the interparticular interactions that determine the rheological behaviors of the cementitious suspensions. A suitable additive can interfere with the interparticle interactions and thus strongly influence the rheology of the fresh cement-based materials [1-4]. As a result, even small fluctuation in the water content can lead to major changes of the fresh concrete behaviors, which are reflected in the changed flowability, different aeration behavior and especially in decreasing mixture stability and segregation. For example, the self-compacting or self-consolidating concrete (SCC) can be adapted very precisely to the growing requirements. This associates with a high sensitivity of the SCC to fluctuations in the quality of the raw materials e.g. admixtures. The setting window of these SCC is very narrow and the handling therefore complicated and difficult. The robustness of the SCC is significantly lower in comparison to conventional concretes, which means that the acceptance in industrial practice is quite moderate [3, 5-7]. A similar situation can be found with fiber concrete and high-strength concrete as well.

Over time, many empirical methods have been developed to determine the properties of fresh mortar and fresh concrete. In addition to the tests on hardened mortars and concretes in the laboratory, tests on plastic state of materials are carried out [8, 9]. These common empirical measurements give robust but limited information on the rheological properties of the mixtures.

The behavior of highly mobile and self-compacting concretes and mortars at the fresh state is much more complex than those of normal slump concretes. Hence, a considerable number of studies concerning the rheological properties of cement-based mixtures were done [3,10-17]. The knowledge of flow mechanics and rheology is very important not only in the area of mortar and concrete development, but also for quality control in the construction laboratory or directly at the construction site. This leads to a necessity in the development of rapid and uncomplicated processes for measuring of rheological behaviors of fresh concrete suspensions.

Rheology is the science of the deformation and flow of matter where the relationship exists between stress, strain, rate of strain, and time. The most important rheological parameters are the yield stress, viscosity and thixotropy (Figure1) [18]. The yield stress describes the behavior of a fluid which behaves like an elastic solid body up to a certain load (flow limit, yield stress) and above like a liquid with plastic viscosity. The viscosity describes the toughness of concrete and the thixotropy (and also rheopexy) as a time-dependent reversible process the change of the yield stress and the viscosity as a function of load. For cement-based mixtures all this processes are depending in time and in the shear history.

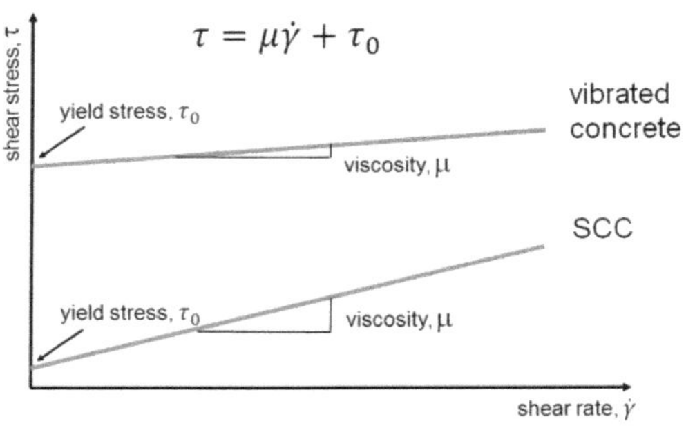

Figure1: Idealized example for rheological curves for vibrated concrete and SCC (according to the Bingham model).

1. Different types of rheometers for different kinds of materials and applications

In order to determine the rheological properties of the fresh concrete, a large number of measuring instruments and measuring systems have been developed by several researchers in the last years [15, 19, 20]. Particularly the particle size distribution and the maximum grain size play an important role for selecting of the right measurement system.

This variety based on different problems appear due to the measurement of the cement-based materials. The common rheometers, as widely used in the food, chemical and pharma industry, are rotational rheometers with a sample size of several milliliters. The maximum particle size of the samples for such rheometers is usually in the range of 0.1 mm.

A further problem of these rotational rheometers is the occurrence of a lubricating film at the interface between the sample and the surface of the sample container and of the probe for non Newtonian fluids. This leads to the measurement of an apparent lower shear stress. On the other side the occurrence of a lubricating film is a key factor for successfully pumpability properties of the materials. Which again highlights the necessary of customized measuring systems.

The main problem is the influence of the flow characteristics and the particle migration and especially particle segregation during the measurement. Within the last 50 years a lot of investigations in the area of prevention of the particle migration during the measuring process in concentrated suspensions were done. This leads in different more or less complicated measurement geometries. In fig. 1 different probe geometries depending on the material mixture and application possibilities are shown.

Depending on the application such as lab application for product development or the field application for quality control different measurement systems which can be combined with different probe geometries are available (fig. 3). One of the important factor for choosing of the right measurement system is the maximum grain size of the cement-based mixture.

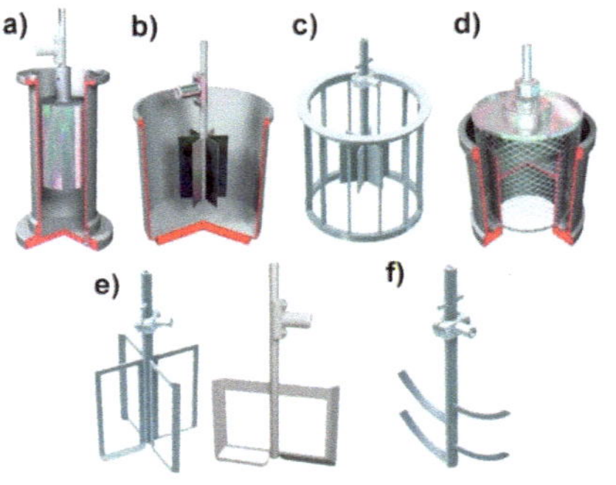

Figure 2: Different kind of probe geometries (Schleibinger Geräte): a) cylinder-cylinder norm probe; b) Vane-probe; c) Vane-probe with antislip roads; d) basket probe; e) probe for mortar and paste; f) probe for mortar and concrete.

Figure 3: Rheology testing equipment from Schleibinger Geräte: Viskomat NT, Viskomat XL, eBT-V and SLIPER (from left to the right).

1.1 Viskomat NT – Rheometer for suspensions up to 2 mm particle size

For the measurement of fine-grained building materials such as cement paste, mortar, fine concrete, plaster with a maximum grain size up to 2 mm Viskomat NT can be used. The Viskomat NT is a versatile rotational and a true speed controlled viscometer driven by a high precision synchronous motor (fig. 4). Each rotation is resolved within 200,000 steps and allows a speed adjustment from 0.001 rpm to 400 rpm in both directions. The torque up ± 500 Nmm is measured by a transducer special developed for this application.

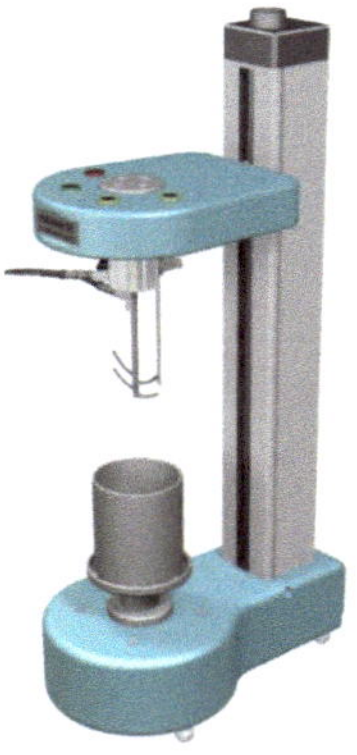

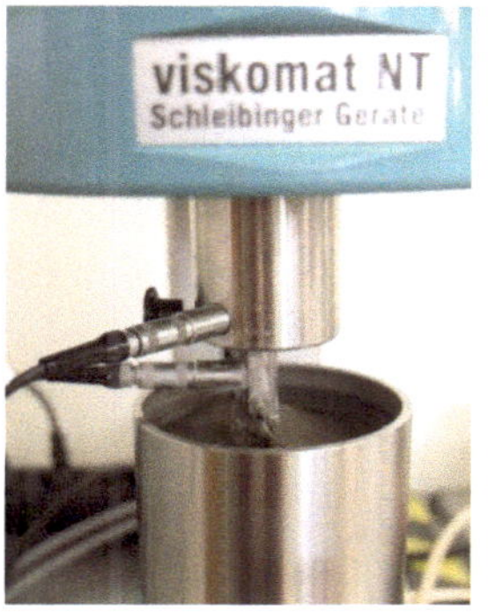

Figure 4: The Viskomat NT - a true speed controlled viscometer.

The measuring system consists of a stationary probe which will be mounted concentrically in a rotation cylindrical sample container. While the sample flows around the probe the shear resistance generates a torque. Torque is transmitted to the measuring probe as a function of the speed of rotation and the flow behavior of the sample.

Different test designs are possible in order to obtain all information about flow curves and rheological parameters. Due to a double walling sample container it is possible to get a temperature dependent workability properties of cement-based materials. The investigation of the stiffening behavior as a function of time and mixing energy as well as effect of concrete admixture and mineral blending agents on workability are easy to realize by simulating the practical application.

1.2 Viskomat XL - Rheometer for suspensions up to 20 mm particle size

The Viskomat XL was specially designed for the measurements of building materials mixtures with the grain size up to 20 mm using the special vane probe geometry. The operation principle of the Viskomat XL is near the same as for the Viskomat NT. Compared to the Viskomat NT with 360 ml of sample volume sample amount of 3 liters is possible.

The rotation speed can be varied between 0.001 rpm and 80 rpm maximum in both directions, clockwise or counter clockwise. The torque in the range of 0 to 1000 Ncm with a resolution of 0.05 Ncm and accuracy better than 0.2 Ncm can be measured.

As well as for the Viskomat NT different measurement profiles like rump or step or rump and step combined can be defined (fig. 5, 6). For special investigations of the inner structure of the materials an oscillating or shear stress controlled mode are possible.

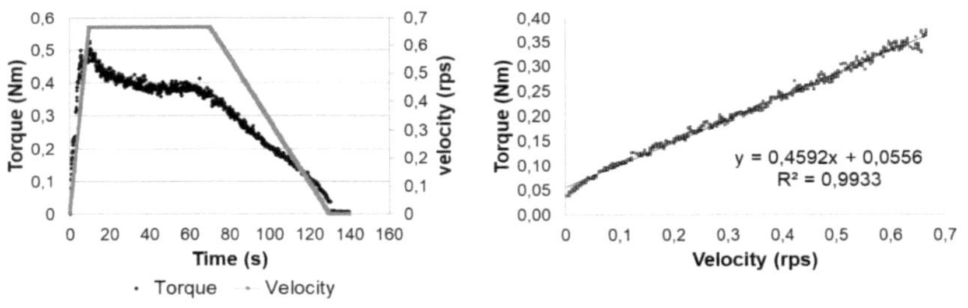

Figure 5: Measurement with Viskomat XL, probe for mortar and concrete (see fig. 2,f): SCC mortar with particle size up to 4 mm measuring profile and measuring data (left) and data evaluation according Bingham model (right).

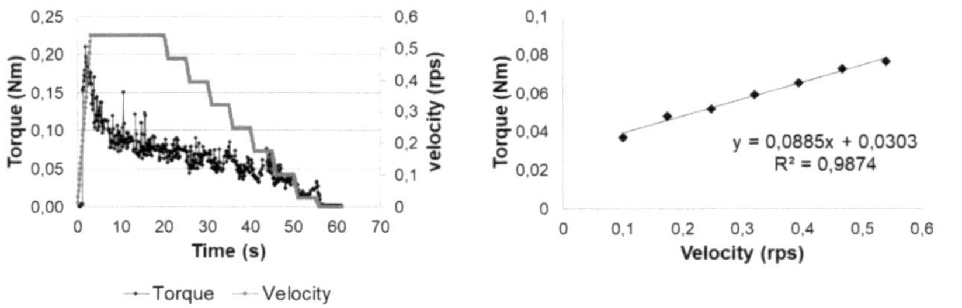

Figure 6: Measurement with Viskomat XL, vane probe: SCC concrete with particle size up to 12 mm measured with step profile (left) and data evaluation according Bingham model (right).

1.3 eBT-V for stiff and self-compacting concretes and suspensions up to 32 mm particle size

Latest development combines two types of concrete rheometers in one (fig. 7Figure). This unit, designed for the field and lab applications, can be operated in two different modes, which are called P-mode and V-mode.

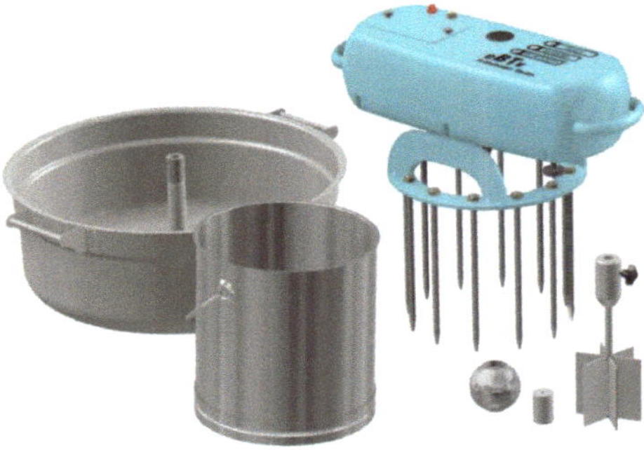

Figure 7: The mobile Rheometer eBT-V from Schleibinger Geräte.

1.3.1 The P Mode – Measuring the force on a small rigid probe shearing the material

By operating the rheometer in P-mode the fresh concrete properties can be investigated due to one single revolution in non-sheared concrete. A measuring probe is attached to the measuring instrument and rotates around a centering rod. The drag-force on the probe is measured independent of the insertion depth. The data of the probe will process into a flow curve taking into account the different angular speeds and the distance of the centering rod. After the evaluation of the measured data according to Bingham model the relative yield stress and the relative viscosity can be determined. Also the stationary force at constant speed, the so called equilibrium shear stress can be calculated [21].

At any time, the measurement is performed in the non-sheared fresh concrete. As a result, problems such as segregation and structural break down during the measurement are greatly minimized. The measurement time is usually less than one minute.

1.3.2 The V-mode – Using the common Vane geometry as approximation of a classical cylinder geometry.

Unfortunately, the drag force on a spherical or cylindrical rigid in a non-Newtonian fluid depending on the viscosity and yield value can't be described in an analytical way yet.

Already in 1936 Russell suggested a so called vane geometry as approximation of the well know cylinder-cylinder geometry for non-Newtonian fluids to eliminate 'the possibility of slipping' [22]. This geometry was commercial available from 1955 by Haake. First Enzler [23] and later Koehler and Fowler [19] suggested this kind of geometry for the use with flowable fresh concrete. In recent years the Vane geometry as a simple and effective measurement method for non-Newtonian fluids has gained in popularity [24]. Once widely used in the food industry [25-27], this probe geometry is increasingly used in the investigation of cement based systems.

Due to the star arrangement of the wings, the sliding on the wall of the probe is prevented. Wall slipping on the container surface is also prevented by the use of a device holder with integrated antislip rods which was developed for this purpose.

The device is fixed on the device holder and the Vane probe is attached on the drive shaft. The measuring container is filled with the sample and the instrument holder with the device mounted on it is placed into container. Due to the built-in variable speed drive, different speed settings are possible to realize. In addition to constant profiles, different step and ramp profiles can be programmed. The rheometer measures the torque on the Vane probe as a function of the speed. The plastic viscosity and the yield stress are calculated from the angular velocity and the measured torque according to Bingham model (fig. 8 and 9). The calculation from torque to shear stress can be done by the mathematical model of Keentok [28] and Barnes [24] or by the model of Koehler and Fowler [19] due to the Reiner-Riwlin equation [29].

For the P-mode and the V-mode simple wireless control via smartphone allows fast and easy operation of the device as well as a fast data transmission via Bluetooth ® and a graphical display of the results. The system is battery driven, water proof and designed for the lab and the building site as well.

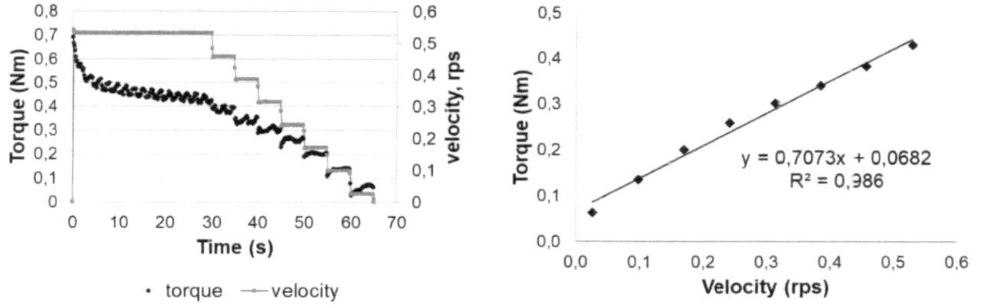

Figure 86: Measurement with eBT-V in Vane mode: SCC mortar with particle size up to 4 mm measured with step profile (left) and data evaluation according Bingham model (right).

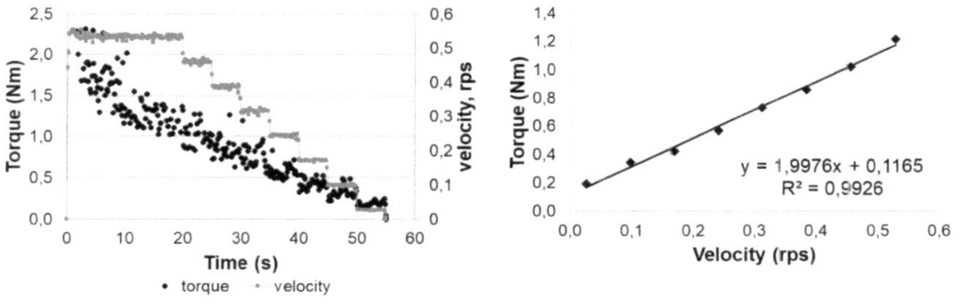

Figure 9: Measurement with eBT-V in Vane mode: SCC concrete with viscosity agent and particle size up to 20 mm measured with step profile (left) and data evaluation according Bingham model (right).

Depending on the mode used, this rheometer is suitable for modern concretes such as SCC and UHPC but also for stiff concretes up to maximum grain size of 32 mm. Additional, this device offers a not negligible practical advantage: The Vane cell and the sample container are easy to clean.

1.4 SLIPER - Measurement of the pumpability of fresh concrete

The flow of suspensions is of fundamental interest and plays an important role in a wide variety of technical applications including paints, cement based mixtures, slurries, and drilling fluids [30]. To build as fast as possible and high as possible concrete is being pumped through pipes to get it to its final destination.

The ability of flow through pipes and hoses by the help of a pump defines the pumpability of concrete [31, 32].

For the pumping process different types of equipment are available. The most important part of the equipment is the pump. When selecting the pump important factors are the maximum pressure, the length of a stroke, and the duration of a stroke. The last two factors are important because they determine the speed of the concrete when it leaves the hopper. The maximum pressure is important because it determines which friction can be overcome with the pumping equipment.

The pipe dimension and particularly diameter is on other factor which influence the speed of the concrete in the pipes. A smaller pipe will result in higher speed, higher speed will result in a higher friction and thus will make the concrete more difficult to pump.

The pumping process is a critical application due to a risk of blockade in the pipe. Problems with the pumpability of concrete result in delays and increased costs. The reason is often depending on the concrete mixture itself. It can happen that the concrete is not suitable for pumping. This problem can be prevented by appropriate measurements of the fresh concrete. Common consistency test like the slump test give no information about the pumpability.

The main challenge is to find a method to quantify the pumpability of different concrete mixes not only in the laboratory, but especially on the construction site direct before use. The SLIPER (SLIding PipE Rheometer), developed by Putzmeister and produced by Schleibinger, allows quick assess of pumping characteristics of concrete and other materials (fig. 10).

"Using the sliding pipe rheometer the time and effort spent on pumping tests can be reduced and thus costs saved. At the same time, this new technology facilitates a quality check, as well as a simple prognosis option with regard to the pumping behavior of concrete and other thick matter," according to concrete technologist and inventor oft he SLIPER, Knut Kasten from Putzmeister.

The SLIPER provides a vertical standing standard pipe with a diameter of 125 mm which is filled with about 7 l of fresh concrete. In the pipe a piston is standing on the ground floor. A pressure sensor is integrated into the top of the piston. If the pipe is sliding downwards, the pressure in the pipe is measured. The speed of the pipe is recorded as well.

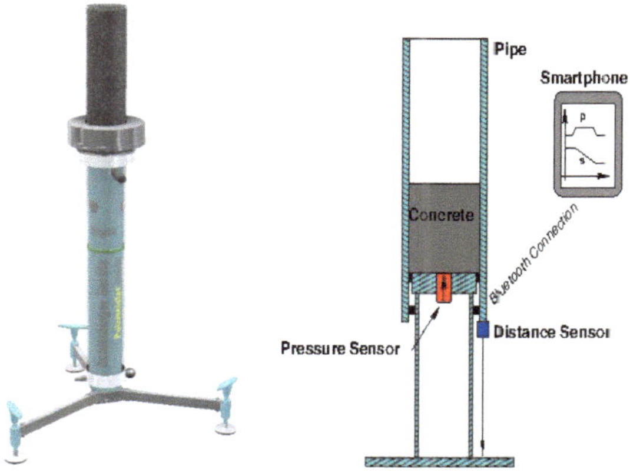

Figure 10: SLIPER – for determining the pumping capacity of concrete.

The properties of the fresh concrete are evaluated by the software App included. With this software the design and parameters for the pump application may be estimated. Therefore, a computational model is used which calculates the expected pressure loss in the concrete pump. The system is portable, robust, battery driven and especially designed for the construction site.

An example of measurement results providing by SLIPER is shown in fig. 11. For these measurements a ready-mix concrete for ceiling construction was taken. The concrete was delivered by trucks to the construction side and were taken for measurement immediately before pumping.

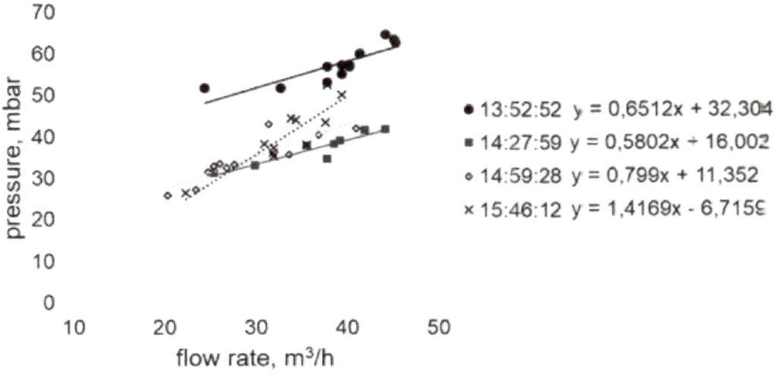

Figure 11: Measurement results of ready-mix concrete for ceiling construction.

The measurement results of ready-mix concrete are showing big variation between the batches and thus quality fluctuations. Based on the results, it is very likely that the quality of the finished product is varying and showing big inhomogeniousities.

Conclusion

The growing demands on the material quality are accompanied by the growing demands on measurement technology. Particularly within the area of concrete production and rheological measurements different types of rheometer for different kinds of materials which records the rheology of the fresh mortar and concrete have been developed. The measurement systems can be specifically adapted to the practice requirement and thus offers new possibilities in the field of concrete development and quality control as well.

Literature

[1] Yang M., Neubauer C.M., Jennings H.M., Interparticle potential and sedimentation behavior of cement suspensions, Adv. Cem. Based Mater. 5 (1), 1997, 1-7.

[2] Uchikawa H., Hanehara S., Sawaki D., Effect of electrostatic and steric repulsive forces of organic admixture on the dispersion of cement particles in fresh cement paste, in Proceedings of the 10th international congress on the chemistry of cement, Vol. 3, 1997.

[3] Hanehara S., Yamada K., Interaction between cement and chemical admixture from the point of cement hydration, adsorption behavior of admixture and paste rheology, Cem. Concr. Res. 29, 1999, 1159-1165.

[4] Flatt R.J., Houst Y.F.A., A simplified view on chemical effects perturbing the action of superplasticisers, Cem. Concr. Res. 31, 2001, 1169-1176.

[5] Lowke D., Schießl P., Robustness of cement suspensions – superplasicizer adsorption, particle separation and interparticle forces, 3rd International RILEM Symposium on Rheology of Cement Suspensions such as Fresh Concrete, Rykjavik, Iceland, 2009.

[6] Wallevik O.H., Kubens S., Müller F., Influence of Cement-Admixture interaction on the Stability of Production Properties of SCC, 5th International RILEM Symposium on Self-Compacting Concrete, Ghent, Belgium, 2007.

[7] Walraven J.C., Takada K., Selbstverdichtender Beton, Zement+Beton, (1) 1999, 23-27.

[8] DIN EN 1015, Methods of test for mortar for masonry.

[9] DIN EN 12350, Testing fresh concrete.

[10] Teubert, J., Measuring the consistency of concrete mortar and its importance to the workalbilty of fresh concrete, Betonwerk + Fertigteil-Technik, Heft 4/81, 1981.

[11] Ferraris C.F., Measurement of the rheological properties of cement paste: a new approach, Role of Admixture in High Performance Concrete, RILEM Internationa Symposium, Proceedings, 1999, 333-342.

[12] Tattersall G.H., Banfill P.F.G., The rheology of fresh concrete, Pitman, 1983, 356pp.

[13] Barnes H.A., Hutton J.F., Walters K., An introduction to rheology, Elsevier, 1989, 199pp.

[14] Banfill P.F.G., A coaxial xylinders viscometer for mortar: design and experimental validation, Rheology of Fresh Cement and Concrete, Spon, 1991, 217-226.

[15] Ferraris C.F., Brower L.E. (editors), Comparison of concrete rheometers: International test at LCPC (Nantes, France) in October 2000, NISTIR 6819, 2001, 147pp.

[16] Banfill P.F.G., The rheology of fresh cement and concrete – a review, Proc. 11th International Cement Chemistry Congress, Durban, 2003.

[17] Banfill P.F.G., The rheology of fresh mortar – a review, Proceedings of 6th Brazilian and 1st International Symposium on Mortar Technology, Florianopolis, Brazil, 2005, 73-82.

[18] DIN 1342-1: 2003-11, Viscosity – Part 1, Rheological concepts.

[19] Koehler E.P., Fowler D.W, Summary of Concrete Workability Test Methods. International Center for Aggregates Research, University of Texas at Austin, 2003.

[20] Greim M., Teubert O., Moderne Messmethoden zur zielgerichteten Entwicklung und Überprüfung der Verarbeitungseigenschaften von Selbstverdichtenden Beton.

[21] Fleischmann F., Kusterle W., A new concrete rheometer for the assessment of the rheological properties of Self-Compacting Concrete, Proceedings SCC 2013, 2013.

[22] Russell J.L., Studies in Thixotropie Gelation, II – The Coagulation of Clay Suspensions, Proc. Roy. Soc. London, Series A, 154, 1936, 550-560.

[23] Enzler R., Lembke E. et. al., Device for testing unset concrete and mortar, Patent CH19910002526 and US5541855, 1991.

[24] Barnes H.A., Nquyen Q.D., Rotating vane rheometry – a review, J. Non-Newtonian Fluid Mech. 98, 2001, 1-14.

[25] ASTM D2573-01, Standard Test Method for Field Vane Shear Test in Cohesive Soil.

[26] Steffe J.F., Rheological Methods in Food Process Engineering, Freeman Press, East Lancing, 1992.

[27] Glenn III T.A., Keener K.M., Daubert C.R., A mixer viscometry approach to use vane tools as steady shear rheologcal attachments, Appl. Rheol. 10 (2), 2000, 80-89.

[28] Keentok M., The measurement of the yield stress of liquids, Rheol. Acta 21, 1982, 325-332.

[29] Reiner M., Deformation and flow. An elementary introduction to theoretical rheology. H.K. Lewis & Co. Limited, Great Britain.

[30] Larson R.G., The structure and rheology of complex fluids, Oxford University Press, New York, 1999.

[31] ACI 1998, Jolin et al, 2006.

[32] Vikan H., Jacobsen S., Influence of rheology on the pumpability of mortar, SINTEF Building and Infrastruction, COIN Project report 21-2010.